ANSWERS TO END-OF-CHAPTER REVIEW QUESTIONS

for

STARR AND TAGGART'S BIOLOGY: THE UNITY AND DIVERSITY OF LIFE

Eighth Edition

David J. Cotter
Georgia College and State University

Wadsworth Publishing Company
I(T)P® An International Thomson Publishing Company

Belmont, CA • Albany, NY • Bonn • Boston • Cincinnati • Detroit
Johannesburg • London • Madrid • Melbourne • Mexico City • New York
Paris • Singapore • Tokyo • Toronto • Washington

Printed in the United States of America
1 2 3 4 5 6 7 8 9 10

For more information, contact Wadsworth Publishing Company, 10 Davis Drive,
Belmont, CA 94002, or electronically at http://www.wadsworth.com/biology

International Thomson Publishing
    Europe
Berkshire House 168-173
High Holborn
London, WC1V 7AA, England

Thomas Nelson Australia
102 Dodds Street
South Melbourne 3205
Victoria, Australia

Nelson Canada
1120 Birchmount Road
Scarborough, Ontario
Canada M1K 5G4

International Thomson Publishing
    GmbH
Königswinterer Strasse 418
53227 Bonn, Germany

International Thomson Editores
Campos Eliseos 385, Piso 7
Col. Polanco
11560 México D.F. México

International Thomson Publishing Asia
221 Henderson Road
#05-10 Henderson Building
Singapore 0315

International Thomson Publishing Japan
Hirakawacho Kyowa Building, 3F
2-2-1 Hirakawacho
Chiyoda-ku, Tokyo 102, Japan

International Thomson Publishing
    Southern Africa
Building 18, Constantia Park
240 Old Pretoria Road
Halfway House, 1685 South Africa

ISBN 0-534-53011-7

TABLE OF CONTENTS

****

PREFACE

This supplement provides answers to the end-of-chapter questions from Starr and Taggart's Biology: *The Unity and Diversity of Life,* Eighth Edition. The questions are meant to help students think critically about the material presented in the book and can serve as a review.

The answers are concise and to the point. They may be direct quotes or paraphrases of the material in the main text, but sometimes answers depart from the material presented in the text. This is done to show students that biology is a broad field that cannot be covered completely in one introductory course, and it is hoped that students will be motivated to read other material for a deeper understanding of and appreciation for biology as one of the most interesting areas of scientific inquiry.

UNIT I INTRODUCTION

CHAPTER 1

CONCEPTS AND METHODS IN BIOLOGY

1. *Why is it difficult to formulate a simple definition of life?* C.1 *For this and subsequent chapters, italics after the review questions identify where you can find the answers to review questions.)They include section numbers and C1 (for Chapter Introduction).* Life is difficult to define as a single entity because it involves a history that encompasses several billion years. The process of evolution (change of living things through time) is part of the concept of life. Life is too complex to define in one simple statement. There is no single quality or feature that is diagnostic of living creatures. To define life, the characteristics of living things are usually listed and compared to inanimate objects.

2. *What is the molecule of inheritance? What is so important about the flow of events from DNA to RNA to protein.* 1.1 The molecule of inheritance is a molecule of deoxyribonucleic acid (DNA). The flow of events from the exact replication of the information contained in the DNA code to its transcription in a specific protein is the key to how DNA controls living processes. It is through this flow of information that a DNA molecule fulfills its destiny to control all facets of life. The protein specified by DNA may be an important structural molecule such as collagen, a transport protein such as hemoglobin or a functional protein such as an enzyme. Faulty instructions or disturbances in the flow of information from DNA to protein may result in genetic diseases such as cystic fibrosis, sickle cell anemia or cancer.

3. *Write out simple definitions of the following terms:* 1.1.
   a. *Cell*   *c.* *Metabolism*   *e.* *ATP*
   b. *Energy* *d.* *Photosynthesis*   *f.* *Aerobic respiration*

(a) Cell: the smallest unit of life that can survive and reproduce on its own and follow the instruction in its DNA code; (b) Energy: the ability to do work; (c) Metabolism: the sum total of all the chemical and physical processes associated with life that control the transfer of energy within cells; (d) Photosynthesis: the trapping and conversion of solar energy by chlorophyll into chemical energy that that can be used by cells; (e) ATP: a molecule of adenosine triphosphate is the energy carrier of cells and is responsible for the transfer of energy involved in metabolic processes; (f) Aerobic respiration: the major energy-releasing metabolic pathway that generates ATP and uses oxygen as its final hydrogen acceptor.

4. *By what mechanisms do organisms sense changes in their surroundings.* 1.1 Organisms have receptors that sense changes in their surroundings. Changes in light-intensity, temperature, pressure, or chemical concentration in the environment are called stimuli. Organisms make controlled compensatory responses to them.

5. *List the shared characteristics of life.* 1.3 A partial list of the characteristics of living things would include: a cellular organization, metabolic processes that allow energy transfers, conversion of inorganic chemicals into living protoplasm, response to changes in the environment, ability to reproduce, grow and develop under the directions of the heritable instructions contained in DNA, and changes that occur over time resulting in evolution.

6. *Study Figure 1.4. Then, on your own, arrange and define the levels of biological organization.* 1.2 Subatomic particles--> atoms-->molecules-->organelles-->cells-->tissues-->organs--> organ systems-->(unicellular organisms)-->multicellular organisms-->populations-->(species)-->communities--> ecosystems-->(biomes)-->biosphere. (Items in parentheses added to the list are presented in the book.) This sequence traces organization in nature from the simple to the complex.

Subatomic particles are neutrons, protons, and electrons, the basic components of atoms. Atoms are the basic building blocks of molecules. They are the smallest indivisible unit of an element that can enter into a chemical reaction. Molecules are two or more atoms joined by a covalent bond. Organelles are distinct structures found in a cell's cytoplasm that perform specific cellular functions. Cells are the structural and functional units of all living things. They are the basic building blocks of life. Tissues are groups of similar cells that perform the same functions. Organs are discrete structures composed of two or more tissues that are integrated to perform one or more functions. Organ systems consist of organs throughout the body that share the same function such as the reproductive, endocrine, or nervous systems. Unicellular organisms are one-celled living creatures such as bacteria, protozoa, algae, etc. Multicellular organisms are living beings that contain more than one cell. They include such diverse forms of life as animals, plants, and fungi. Populations are groups of organisms that inhabit a common area where they are able to interbreed. Populations are the functional units of evolution. Species are a group of organisms that share the same gene pool, that is they interbreed and share the same genetic features. Communities contain all the populations that inhabit the same area. Ecosystems are the functional unit in nature. They consist of all the living organisms and raw materials within a given area. Biomes consist of large mature communities that inhabit large areas and are characterized by characteristic plants and animals as determined by shared environmental conditions. Biosphere is the sum total of all living organisms on earth and their interactions with the physical and chemical environment.

7. *Study Figure 1.5. Then, on your own, make a sketch of the one-way flow of energy and the cycling of materials through the biosphere. To the side of the sketch, write out definitions of the producer, consumer, and decomposer organisms.* 1.2 The ultimate source of energy in the biosphere is the sun. Sunlight reaches the earth, where it is captured by green plants and converted into food. The energy locked in food can be used by the plant to drive various biological reactions, or it may be transferred to animals that feed upon it. The energy may be transferred from one organism to another in a process known as energy flow. Once energy has been used, it is dissipated as low grade heat, and more energy has to be resupplied from outside of the system. Energy is captured by

producers and is transferred or flows from producers to consumers. When the producers or consumers die, the energy locked in their bodies is transferred or flows to the decomposer. A producer is any organism that captures energy from sun (photosynthesizers) or from chemical reactions (chemosynthesizers). A consumer is an organism that feeds upon producers and utilizes the energy stored by them. A decomposer is any organism that uses the energy stored in producers or consumers after they have died.

Essential materials such as chemical elements are present on earth. Those associated with living organisms are part of a biogeochemical cycle. They are not broken down but retain their entities as chemical elements regardless of whether they are involved in biological, chemical, or geological reactions. They move from one chemical compound to another to recycle continuously.

8.   *Each kind of organism is called a separate species. What are the two components of its name? List the six kingdoms of species and name some of their general characteristics.* 1.3 The scientific name for an organism as a binomial includes the genus and the species such as <u>Homo</u> <u>Sapiens</u>. The six kingdoms of life include the following: (1) Archaebacteria, ancient bacteria are simple prokaryotes that live in extreme conditions similar to those that existed when life first appeared. (2) Eubacteria, true bacteria that are ubiquitous, that is they are found everywhere. They are the familiar bacteria that function as pathogens or decomposers. (3) Protista are mainly unicellular although some are multicellular. They are the first eukaryotes and are much more complex than the prokaryotes, possess a nucleus and a variety of cellular organelles. Many are single-celled algae or protozoa. (4) Fungi are heterotrophic organisms that function as decomposers or parasites. Most are multicellular and include the mushrooms. They secrete enzymes that digest food outside their bodies and then absorb the dissolved food. (5) Plants are usually multicellular producers that carry on photosynthesis. There are some parasitic plants. 6) Animals are familiar multicelled consumers that feed upon plants or other animals. They generally move about in contrast to plants that usually remain in one place.

9. *Define mutation and adaptive trait.* 1.4 A mutation is any inheritable change in a gene. Mutations provide an array of characteristics that are subjected to natural selection to produce the current survivors of evolution. An adaptive trait is an inherited characteristic that permits an organism to survive and reproduce under a given set of environmental conditions. Traits that are non-adaptive reduce an organism's chance to survive and reproduce under a given set of environmental conditions

10. *Explain the connection between mutation and the immense diversity of life?* 1.4 DNA stands for deoxyribonucleic acid, the most important chemical in the world. It is the gene, or living blueprint, and as such is responsible for all the diversity that we see in living organisms. A mutation is an inheritable change in a DNA molecule. The genes of all the organisms alive today have withstood the process of natural selection throughout their evolution. Any change produced now is likely to be one that had been tried in the past and found wanting; therefore, it would still have less chance to survive and reproduce than the preexisting genes that have withstood the selection process. A favorable mutation is rare but it confers an advantage in survival and reproduction to the organism that expresses it. Therefore, when the extremely rare favorable mutation arises it has a chance to spread through a population. Mutations are responsible for the production of new genes and are the driving force for evolution. The differences we see in living organisms represent biological diversity. These differences are the phenotypic expression of different genes produced by random mutations that arise through time.

11. *Write out simple definitions of evolution, artificial selection and natural selection.* 1.4 Evolution is the change in a group of related organisms over time. It involves a change in gene frequencies in a population through time. Artificial selection involves the selection of different traits in an artificial environment as opposed to natural selection which involves the natural environment. In artificial selection man sets forth the criteria for selection, which, for example, results in the wide variety seen in the breeds of domestic dogs and other animals. Natural selection results in differential survival and reproduction among members of a population that differ in one or more traits.

12. *Define and distinguish between*: 1.5
   a. *Hypothesis or speculation and scientific theory*
   b. *Observational test and experimental test*
   c. *Inductive and deductive logic*

A hypothesis or speculation is a potential explanation for a natural phenomenon. A hypothesis is usually stated as a question that can be answered by conducting tests or experiments. It is one of the first steps in the scientific method. A scientific theory is a generalization that is formulated after a series of further observations or experiments that enables accurate prediction of the results of further tests or observations.

Observational test simply means making further observations about the phenomenon under study to see if the observation supports the hypothesis or generalizations that were made after the original observations. Experimental tests involve establishing two groups of subjects, a control group and an experimental group. The experimental group is exposed to a variation in a single variable that is being tested. The control group is identical to the experimental group in all respects except for the variable being tested. The difference in the response of the two groups is due to the difference in the variable.

Deductive logic proceeds from the general to the specific. It involves two generalizations called a major premise and a minor premise and a specific conclusion based upon the relationships of the two premises. For example: All men are mortal; Socrates is a man; therefore Socrates is mortal. On the other hand, inductive logic proceeds from the specific from the general. A series of specific observations or tests allow the development of a generalization (scientific principle, theory, law).

13. *With respect to experimental tests, define variable, control group, and experimental group. What is meant by the statement that scientific experiments are based on an assumption of cause and effect.* 1.5, 1.6 A variable in an experimental test is the factor that differs between the experimental and the control group. It could be the amount of light, the presence or absence of a chemical, temperature, diet, genetic background, etc.

The control group is the group used to evaluate the effects of a variable on the experimental group. The experimental group differs from the control group by only the one factor being tested. Science assumes that there is one or more underlying causes for a specific effect that can be observed in the natural world. Often this may be expressed as an if ... then statement: i f a certain cause exists, then a certain effect is produced. Science is based upon predictions that certain conditions result from predictable events.

14. *What is a sampling error?* 1.6 Sampling error is based upon random variation in a population. If a sample is not large enough or is not a true representation of a population then assumption made using an inadequate or inappropriate sample may contain sampling error. The validation of statistical analysis depends upon a large enough sample.

CHAPTER 2

CHEMICAL FOUNDATIONS FOR CELLS

1.   *Name the four elements (and their symbols) that make up more than 95 percent of the body weight of all organisms.* 2.1 The four elements that make up more than 95 percent of the body weight of all organisms are: Oxygen (O), Carbon (C), Hydrogen (H), and Nitrogen (N).

2. *Define isotope, and describe how radioisotopes are used either in radiometric dating or as tracers.* 2.1, 2.2 The word isotope means equal place. It refers to the periodic table of elements. (Appendix VI displays a periodic table of elements). All isotopes of an element fit in the same place in the periodic table. For example, there are three isotopes of hydrogen. H^1 (Hydrogen) has one proton and an atomic weight of 1. H^2 (deuterium) has one proton with an atomic weight of 2. H^3 (tritium) has one proton with an atomic weight of 3. All isotopes of hydrogen have one proton and therefore fit into the same place of the periodic table and all behave the same chemically and biologically. The difference in

the isotopes is physical, based upon the number of neutrons it possesses. Hydrogen has no neutrons, deuterium has one neutron and tritium has two neutrons. The ratio of neutrons to protons in tritium is unstable and it gives off weak beta radiation and is radioactive. The other two forms of hydrogen are stable but can be distinguished by a mass spectrograph that measures the difference in atomic weight.

Each radioactive isotope gives off radiation (that is why it is called radioactive) as it undergoes radioactive decay. The imbalance in the neutron/proton ratio makes for differences in amount of instability of the atom's nucleus and therefore rate of decay varies for each isotope. For example, Cobalt 51 has a half-life (length of time it takes for 50 percent of a radioactive sample to decay) of 5.2 years. Carbon 14-5,730 years. Phosphorus 32-14 days, Iodine 131-8.1 days. Some half lives are measured in fractions of a second while others in billions of years. The difference in half-lives enable the determination of the age of samples that contain these isotopes. This dating of samples is known as radiometric dating. For example, the dating of the radioactive carbon found in the Dead Sea Scrolls found that they were approximately 2,000 years old plus or minus 100 years. In other words, they were authentic in terms of age.

Since radioactive isotopes behave the same chemically and biologically, it is possible to use radioactive isotopes to trace biological reactions. For example, if radioactive Iodine is administered to a patient, the majority of it will become incorporated in the thyroid gland. Thus a radioactive cocktail containing radioactive iodine could be a way to irradiate excess thyroid tissue and reduce its size without an operation. Likewise, radioactive tritium can be administered to cells to understand how genes are replicated. Radioactive iron could be administered to a patient to determine blood volume by isotope dilution (after an amount has been administered and given a chance to thoroughly mix a small sample of blood will contain some radioactive iron). It is then a simple matter to determine how much the radioactive iron has been diluted to calculate the volume of blood in the patient.

3.   *How many electrons can occupy each orbital around an atomic nucleus? Using the shell model, explain how the orbitals available*

8

to electrons are distributed in an atom. 2.3 One or at the most, two electrons may occupy an orbital. The first shell has a single spherical orbital which may hold one or two electrons. The second shell or energy level has four orbitals each of which can hold a maximum of two electrons. The third shell or energy level also has four orbitals with a maximum of two electrons/orbital. Other shells contain additional orbitals that hold sufficient electrons to balance the number of protons in the nucleus.

4. *Distinguish between:*
    a. *Ionic and hydrogen bonds* 2.4
    b. *Polar and nonpolar covalent bonds* 2.4
    c. *Hydrophilic and hydrophobic interactions* 2.5

In an ionic bond, a positive ion and a negative ion are linked together by the mutual attraction of opposite charges. In a nonpolar covalent bond, two atoms are bound together by sharing an electron equally. A hydrogen bond is an example of a polar ionic bond. A hydrogen atom in compounds such as water has a slight positive charge and exhibits an attraction to electron negative atoms that are already participating in a polar covalent bond. This attraction is a hydrogen bond

In polar covalent bonds, two atoms of two different elements with different number of protons share pairs of electrons. The atom with the greater number of protons will exert a greater attraction for the pair of electrons, thus ending with a greater negative charge and is described as the electronegative end of the molecule. The other end is electropositive. In the water molecule, the oxygen atom has more protons and attracts the electrons more and is therefore electronegative. The water molecule exhibits polarity. On the other hand, a hydrogen molecule (H_2) is nonpolar because there is no difference in the attraction of either of the hydrogen for the shared electrons so that there is no difference in polarity established, hence a nonpolar covalent molecule.

If a substance is polar it is attracted to water molecules and is called hydrophilic (water loving). Hydrophobic (water fearing) molecules are nonpolar. Hydrophobic molecules tend to be rejected by water molecules that attract one another and form hydrogen bonds.

5. *If a water molecule has no net charge, then why does it attract polar molecules and repel nonpolar ones?* 2.5 The existence of the slight polarity is exhibited because of the greater attraction of the oxygen's protons than the hydrogen's protons. The oxygen end of the molecule is electronegative and the hydrogen end is electropositive. These differences in charges attract the hydrogen end of one water molecule to the oxygen end of another molecule producing a weak hydrogen bond. These hydrogen bonds attract other polar molecules. Substances that readily bond with water are called hydrophilic. Water's polarity repels nonpolar substances known as hydrophobic molecules.

6. *Define acid and base, then describe the behavior of a weak acid in solution having a high or low pH value.* 2.6 An acid is a substance that releases a proton into a solution. A base is a substance that accepts or combines with a hydrogen ion in solution. Acids may be defined as weak acids or strong acids depending on the pH values. A pH of 7 is neutral. Values below 7 are acidic. The lower the number, the stronger the acid. Strong acids totally give up hydrogen ions when they dissociate in water. Weak acids are less likely to give up hydrogen ions. A weak acid may even accept hydrogen ions and behave as a base.

CHAPTER 3

CARBON COMPOUNDS IN CELLS

1. *Define organic compound. Name the type of chemical bond that predominates in the backbone of such a compound.* 3.1 An organic compound is a compound synthesized by living cells around carbon molecules in straight chains or rings. The most common bonds in the backbone of organic compounds are single covalent bonds between carbon atoms.

2. *Name the "molecules of life." Do they break apart most easily at their hydrocarbon portion or at functional groups.* 3.1, 3.2 The "molecules of life" are synthesized by living cells and include

carbohydrates, fats, proteins and nucleic acids. Reaction of these organic compounds primarily involves enzyme-mediated transfer of functional groups or electrons, rearrangement of internal bonds and combining (condensation) or splitting of molecules.

3. *Select one of the carbohydrates, lipids, proteins, or nucleic acids described in this chapter. Speculate on how its functional groups as the bonds between carbon atoms in its backbone contribute to its final shape and function.* 3.4, 3.5, 3.6-3.7, or 3.8 Starch is a large complex carbohydrate polymer composed of many monomers of glucose. The angles of the covalent bonds joining each individual glucose monomer shape the large polymer into a spiral coil. These starch chains are not very stable and those starch chains with branches are even less stable. They are arranged so that hydroxyl (-OH) groups are exposed for easy enzyme access. As a large molecule used to store carbohydrates, starch is relatively immobile, but it can be broken down enzymatically to simpler compounds in order to distribute the food in the organism.

Triglycerides are neutral fats composed of three fatty acids attached to a backbone of glyceride. They are the richest energy compound available. When compared to carbohydrates the most notable difference is the small amount of oxygen in fats. Much more oxygen has to be added to a fat than a carbohydrate to break it down to carbon dioxide and water. In the process, more chemical bonds are broken and two and a quarter times as much energy is released in a gram of fat then a gram of carbohydrate. Animal fats are solid at room temperature and layers of fat in the body provide insulation.

Proteins exhibit four levels of structure which affect the functioning of the molecule. For example, a change in the primary structure of a protein by altering the sequence of the amino acids in the hemoglobin molecule accounts for the difference in normal hemoglobin and sickle-cell hemoglobin. Sulfhydryl and hydrogen bonds are important in controlling the quaternary structure of proteins and contribute to the shape of the active site of a protein substrate that is acted upon by an enzyme.

A nucleic acid is composed of nucleotides which contain a phosphate group, a pentose sugar and a nitrogenous base. The backbone of the nucleic acid is constructed so that the sugar unit is covalently bonded to the phosphate group of adjacent nucleotides. The DNA molecule is a long double-stranded chain of nucleotides that form a double helix. Hydrogen bonding between nucleotide bases hold the two strands of DNA together. These hydrogen bonds can be separated during replication (synthesizing another DNA molecule) or transcription (synthesizing a RNA molecule). The sequence of nucleotides contain the genetic code. The exact process of converting the DNA code into a specific protein will be covered in detail in Chapter 14, but the structure of the DNA molecule and the sequence of its functional groups is at the heart of the explanation of the chemical nature of a gene.

4. *Which category includes all of the other items listed?* 3.5
   a. *Triglyceride*   c. *Wax*   e. *Lipid*
   b. *Fatty acid*   d. *Sterol*   f. *Phospholipid*

Lipid is the general name for all fats and includes phospholipids, sterols, triglycerides, fatty acids, neutral fats, oils, and waxes.

5. *Explain how a hemoglobin molecule's three-dimensional shape arises, starting with its primary structure.* 3.6, 3.7 The hemoglobin molecule is a complex protein composed of a central heme unit with four different globulin molecules, two called alpha chains and the other two, beta chains. The primary structure is controlled by the specific sequence of amino acids. As you will discoverer in chapter 14, a single gene mutation results in the substitution of the amino acid valine for the normal glutamate in the sixth amino acid in the beta chain. Glutamate has a negative charge while valine has no charge. This substitution renders basic changes in the ability of the defective hemoglobin molecule to transport oxygen.

Each individual amino acid has its own shape depending upon the number and arrangement of the atoms used to make up the amino acid (tryptophan is much larger than glycine). The oxygen and other atoms of specific amino acids in the sequence affect the pattern of hydrogen bonding between different amino acids along the chain. In addition, the R groups in the sequence interact and determine how the chain bends and twists into its three-dimensional shape. The common shapes produced are helical or spiral chains or a sheetlike array. Often the helically coiled chains will become further folded when one R group interacts with another. Further structural modification occurs when two or more polypeptide chains interact to form globular complexes.

The four levels of protein structure are as follows: primary structure—the particular sequence of amino acids; secondary structure—the pattern produced by the hydrogen bonding resulting in a spiral (helix) or pleated sheets; tertiary structure—folding of the proteins because of interaction of its R groups with disulfide bonds and other bonds; quaternary structure—interaction between two polypeptides resulting in the globular shape of any of the giant protein molecules. The structure of protein molecules may add strength and rigidity. Spatial arrangements are particularly important in proteins that function as enzymes. When the three-dimensional shape of a protein is disrupted, the protein is said to be denatured, and it unwinds and changes shape.

The primary structure influences the protein's structure in two major ways. First, it controls the hydrogen bonding and secondly it arranges the positions of each amino acid to enable them to react and bend to reach a stable three dimensional shape. The secondary structure of a protein depends upon whether the molecule is coiled or extended depending upon the patterns of hydrogen bonding along the polypeptide chain. Coiled chains fold into distinctive shapes when one R group interacts with another some distance away or with part of the polypeptide's backbone or substances within the cell. The folding that arises through hydrogen and disulfide bonds and interactions between the R groups of a polypeptide chain represents the tertiary structure of the molecule. The quarternary structure of some proteins are forms of two or more protein chains that are folded together as the four globulin chains in the hemoglobin molecule.

CHAPTER 4

CELL STRUCTURE AND FUNCTION

1. *Label the organelles in this diagram of a plant cell.* 4.3 The organelles in the diagram of a typical plant cell, starting at the one o'clock position and proceeding clockwise are as follows: central vacuole, nuclear envelope, DNA and nucleoplasm, nucleolus, endoplasmic reticulum, microtubules, plasma membranes, mitochondrion, chloroplast, microfilaments, vesicle, and Golgi body.

2. *Label the organelles in this diagram of an animal cell.* 4.3 The organelles in the diagram of a typical plant cell starting at the one o'clock position and proceeding clockwise are as follows: cell wall, plasma membrane, central vacuole, microfilaments, vesicle, Golgi bodies, smooth endosplasmic reticulum, rough endoplasmic reticulum (ribosomes attached), nucleus (nuclear envelope, nucleolus, nucleoplasm), mitochondrion, and chloroplast.

3. *State the three key points of the cell theory.* 4.2 All living things are made up of cells. Cells are the structural and functional units of life. All cells arise from preexisting cells.

4. *Describe three features that all cells have in common. After reviewing table 4.2, write a paragraph on the key differences between prokaryotic and eukaryotic cells.* 4.1, 4.3, 4.11 The three features all cells have in common are: (1) a thin outer plasma membrane that controls what goes in and out of the cell, (2) a DNA-containing region that exerts control over the cell and supplies hereditary information, and (3) cytoplasm that is enclosed by the plasma membrane and contains organelles that help the cell carry out its function.

Prokaryotic cells are characteristic of the archaebacteria and eubacteria. They are much simpler, smaller, and more primitive than eukaryotic cells found in protista, fungi, plants, and animals. Prokaryotic cells do not have a separate nucleus. Their DNA usually consists of one long DNA molecule and sometimes small fragments of DNA in structures called plasmids. The prokaryotes also have RNA and ribosomes to translate the genetic code into proteins. Prokaryotes have both a cell wall and a plasma membrane. Some have photosynthetic pigments and some have locomotive organelles such as flagella.

The most obvious feature of Table 4.1, which describes the cellular components of prokaryotic and eukaryotic cells, is the large number of organelles that are not found in the prokaryotes. Prokaryotic cells lack a nucleus, nucleolus, the endoplasmic reticulum, Golgi body, lysosomes, chloroplasts, central vacuole, cytoskeleton, and a complex flagellum or cilium. The reason that most of these features are lacking may be that the prokaryotes are simple and primitive organisms, and these organelles had not evolved in the Monerans. The prokaryotes do have a cell wall, plasma membrane, DNA and RNA molecules, ribosomes, and photosynthetic pigments and are fully capable of functioning at a lower level of complexity.

The eukaroyote is much more complex. The major feature is the presence of a nucleus with a nuclear envelope, a fluid nucleoplasm, and one or more nucleoli. All of the eukaryotes except the animals have cells with a protective cell wall and all have a plasma membrane that controls the ingress and egress from the cells' interior. The cytoplasm contains a wide variety of organelles that depend upon the species or the type of tissue considered. Some of the most common along with their typical functions are as follows:

Mitochondria - production of energy
Chloroplast - capture of solar energy
Ribosome - protein synthesis
Endoplasmic reticulum - internal transport in the cells along with lipid synthesis
Golgi body - production of vesicles for internal transport and export from the cells
Lysosomes- intracellular digestion
Central vacuole - storage
Cilia and flagella - cellular motion
Cytoskeleton - cell shape and motion

5. *Suppose you want to observe the three-dimensional surface of an insect's eye. Would you benefit most using a compound light microscope, transmission electron microscope, or scanning electron microscope?* 4.2 The scanning electron microscope is ideal for

studying surface features. While it does not magnify to the extent of high-resolution transmission microscopes, it does provide a three-dimensional image, ideal for studying surface features.

6. *Briefly characterize the structure and function of the nucleus, the nuclear envelope, and the nucleolus.* 4.4 The nucleus is the diagnostic feature that distinguishes eukaryotic cells from prokaryotic cells. The nucleus is separated by the nuclear envelope from the cytoplasm that forms the remainder of the cell. The nuclear envelope has pores that allow substances to pass across it. Inside the nuclear membrane is the liquid nucleoplasm, the nucleolus and chromatin material. The nucleoplasm has materials dissolved in it and its fluid nature allows transport throughout the nucleus. The number of nucleoli found in a nucleus varies with age. These are sites where proteins and RNA subunits for ribosomes are assembled and later transported to the cytoplasm. . The chromatin material will condense into rod-like bodies (chromosomes) at the time of cell division. In the normal non-dividing cell the chromosomes are dispersed into threads that consists of proteins and DNA molecules.

7. *Define chromosome and chromatin. Do chromosomes always have the same appearance during a cell's life?* 4.4 Chromosome means colored body. It is a molecule of DNA with associated proteins that carries the genetic code. Human nuclei have 24 pairs of chromosomes in the nucleus of each body cell. The name chromatin refers to its colored threadlike material when the chromosome is dispersed in a relaxed state. The chromatin will shorten and coil into a small rod-shaped body during cell division to facilitate movement.

8. *Which organelles are part of the cytomembrane system?* 4.5 The cytomembrane system is a series of organelles where lipids are assembled and polypeptide chains are modified into the final form of proteins. The cytomembrane system includes the following: plasma membrane, transport vesicles, lysosomes, Golgi bodies, microtubules, smooth and rough endoplasmic reticulum, cytoplasm and nuclear membrane.

9. *Is this statement true or false? Plant cells have chloroplasts, but not mitochondria. Explain your answer.* 4.6, 4.7 The first part of the statement is essentially true, but the second part is false. Most

plants are green and have chloroplasts where photosynthesis takes place. Cells found in some parts of a plant, such as underground roots and interior of stems may lack chloroplasts. There are a few parasitic plants that lack chloroplasts. All eukaryotic organisms, including plants, have mitochondria and carry on aerobic respiration to liberate the energy available in glucose. Plants carry out photosynthesis in the sunlight but carry on respiration all the time.

10. *What are the functions of the central vacuole in mature, living plant cells?* 4.7 The function of the central vacuole in plant cells may vary depending on the cell. It is one way to enlarge the cell. As the vacuole increases in size, the cell itself enlarges. It also may serve as a storage area for metabolic products, such as ions, sugars, amino acids, wastes, toxins, and pigments.

11. *Define cytoskeleton? How does it aid in cell functioning?* 4.8 The cytoskeleton is a three-dimensional system that gives stability and structure to the cell. It is composed of interlocking microfilaments and microtubules. They function in cell movement and during cell division. They enable organelles to be arranged in appropriate sequence for metabolic pathways.

12. *Are all components of the cytoskeleton permanent?* 4.8, 4.9 The cytoskeleton in all eukaryotic cells consists mainly of microtubules and microfilaments. Intermediate filaments are found in animals cells. All three are assembled from subunits of proteins. Microtubules are composed of tubulin, and some are transitory features used to form spindle fibers that move chromosomes during mitosis. On the other hand, the microtubules found in cilia and flagella are permanent structural features of these organelles. Actin and myosin are permanent functioning microfilaments found in muscle cells. The keratin found in hair cells is derived from intermediate filaments.

13. *What gives rise to the 9+2 microtubular array of cilia and flagella?* 4.9 The centrioles give rise to the microtubular array. After the cilium or flagellum has been formed, the centriole remains attached at its base, just below the plasma membrane.

14. *Cell walls are typical of which organisms: animals, plants, fungi, protistans, or bacteria? Are cell walls solid or porous?* 4.10 Cell walls are found among plants, fungi, bacteria, and protistans, but not in animals. The characteristics of cell walls vary depending on construction, but all are porous to allow entrance and exit of water and other material. In some specialized cells, the cell walls may be impregnated with chemicals which prevent the penetration of water, such as the waxy substance found in the cell walls of cork cells.

15. *In certain plants cells, is a secondary cell wall deposited inside or outside the surface of the primary cell wall?* 4.10 In plants, the secondary cell wall is deposited inside the primary walls. Not all plants develop secondary cell walls.

16. *In multicelled organisms, coordinated interactions depend on linkages and communications between cells. What types of junctions occur between adjacent animal cells? Plant cells?* 4.10 In animal cells there are three types of cell-to-cell junctions: (a) tight junctions, which occur between cells of epithelial tissues, (b) adhering junctions, which are like welded spots between adjacent epithelial cells so that they maintain contact when stretched and, (c) gap junctions, which are small channels that allow instant communication by the flow of material from one cell to the adjacent cell. In plant cells, the linkages occur between cell walls, not plasma membranes. These junctions are called plasmodesmata. There may be as many as 100,000 plasmodesmata penetrating through primary and secondary walls of a single cell to allow rapid communication between cells.

CHAPTER 5

A CLOSER LOOK AT CELL
MEMBRANES

1. *Describe the fluid mosaic model of cell membranes. What imparts fluidity to the membrane? What makes it a mosaic?* 5.1 The fluid mosaic model of a membrane has a lipid bilayer with hydrophilic heads on the outsides of the bilayer and hydrophobic tails between the two rows of hydrophilic heads. There are proteins scattered throughout the membrane. In some cases, the proteins will not penetrate all the way through the bilayer, and other times the proteins will extend beyond the membrane on each side. The fluid feature of the membrane is the result of the nonpolar lipids that have proteins floating in them. The mosaic is the intricate composite of diverse proteins and lipids of the membranes.

2. *Structurally, what do all cell membranes have in common? In what ways do the membranes of different cell types vary?* 5.1 Cell membranes are composed of lipids (phospholipids especially) and proteins. Membrane lipids have hydrophilic heads and hydrophobic tails, and when surrounded by water they assemble spontaneously into a bilayer. All heads are at the two outer faces of a lipid bilayer, and all tails are sandwiched between them. The lipid bilayer is the basic structure of all cell membranes. It acts as a hydrophobic barrier between two fluid regions (either two cytoplasmic regions, or the cytoplasm and the fluid outside the cell). Membrane fluidity arises through rapid movements and packing variations among the individual lipid molecules. Membrane functions are carried out largely by proteins embedded in the bilayer and are weakly bonded to either surface bilayer. Some proteins transport water-soluble substances across the membrane. Some are enzymes. Others are receptors for chemical signals or for specific substances. Additionally, some recognition proteins which are like molecular fingerprints, serve to identify a cell as being of a specific type and belonging to a specific organism.

3. *Distinguish among transport proteins, receptor proteins, recognition proteins, and adhesion proteins.* 5.1 Transport proteins extend through the lipid bilayer of the membrane. They allow water-soluble substances to travel through their interiors. They use energy to pump substances one way through the lipid bilayer. Both receptor and recognition proteins have specific three-dimensional shapes and extend outside the membranes. Receptor proteins act like switches that are turned off and on when specific structures such as hormones bind to them. Recognition proteins function in tissue formation and cell to cell interactions. They are molecular fingerprints that serve as markers to alert the immune system that they are "self" proteins and are therefore not to be attacked by marauding white blood cells. Adhesion proteins help cells of the same type locate and stick to each other and remain in their proper position and may become a type of cell junction.

4. *Define diffusion. Does diffusion occur in response to a solute concentration gradient, an electric gradient, a pressure gradient, or some combination of these?* 5.3 Diffusion is the movement of the molecules of one substance through the molecules of another substance from regions of higher concentration to regions of lower concentration by means of their own kinetic energy. The molecules will move from high to low concentrations until the concentration is equal throughout and an equilibrium is reached. The collision of molecules still occur at random and there is no net movement in any direction. If the temperature of the system increases, the kinetic energy increases and the rate of diffusion increases. The greater the concentration gradient the faster the rate of diffusion. Small molecules will tend to move faster than large molecules and will therefore diffuse faster. Differences in electrical charges of molecules may produce an electrical gradient. The opposite electrical charges will attract each other and contribute the flow of charged particles. Differences in pressures may alter the direction and rate of diffusion. Thus, the rate and direction of diffusion can be controlled whenever a gradient exists in concentration, electrical charge, pressure or any combination of all three. Diffusion results in the dispersion of substances by random collisions of molecules.

5. *Define osmosis.* 5.4 Osmosis is the movement of water across any differentially permeable membrane in response to concentration gradients, pressure gradients, or both. Osmotic movement is dependent upon the concentration of solutes in the water on both

sides of the membrane. Water will move down a concentration gradient, that is, it will move from the side of the membrane with the greater water concentration across the membrane to the other side where the water concentration is lower.

6. *Define hypertonic, hypotonic, and isotonic solutions. Does each term refer to a property inherent in a given type of solution? Or are they used only when comparing one solution to another?* 5.4 The application of etymological derivation of terms is an excellent method to understand biological terminology. Hyper/tonic is derived from hyper=above and tonic=strength. Hypo/tonic=below/strength. Iso/tonic=equal/strength. A hypertonic solution is a stronger solution than the solution found in a cell that is placed in it. In other words, the concentration of solutes in the solution is greater than the concentration of solutes in the cell. If the concentration in the solution outside a cell is stronger, it has a lower concentration of water. The concentration inside the cell is weaker when compared to the surrounding solution and therefore has a greater concentration of water. According to the concentration difference, osmosis will result in water moving from its higher concentration in the cell to the surrounding solution with proportionately less water and more solute. As the cell loses water, its cytoplasm will shrink in a process known as plasmolysis. If, on the other hand the cell is placed in a hypotonic solution, the water will move from the weaker solution into the cell causing the cell to expand and the cell will undergo plasmolopsis. Finally, if a cell is placed in a solution that has the same concentration of solutes as the solution within the cell there would be no concentration gradient. Random movement of water molecules would occur, but there would be no net movement of water in either direction. The system would be at equilibrium and conditions would remain the same.

7. *If all transport proteins shunt substances across cell membranes by changing shape, then how do the passive transporters differ from the active transporters?* 5.5 The chief difference between active and passive transport is that in active transport energy in the form of ATP is used to transport material against a concentration gradient. In passive transport movement is random and higher concentrations allow more molecules of the substance to contact vacant binding sites in the interior of the transport protein. In active transport, ATP improves the fit of the binding site on one

21

side of the membrane. Once the solute molecule becomes bound, the transport protein changes shapes to allow the molecule to become exposed to the other side of the membrane. The binding site then reverts to a less attractive state so that both the solute molecule is released and the site is not attractive to other solute molecules on that side of the membrane.

8. *Define exocytosis and endocytosis. Describe the main features of the three pathways of endocytosis.* 5.3, 5.5 Both exocytosis and endocytosis involve the movement of vesicles inside the cytoplasm. In exocytosis, the vesicles are formed inside the cytoplasm by organelles such as the Golgi body. The vesicle moves through the cytoplasm, comes in contact with the plasma membrane, and fuses with it, and the contents of the vesicle are expelled outside of the plasma membrane. In endocytosis, a portion of the plasma membrane encloses a substance and then pinches off, forming a vesicle that moves around inside the cell. Exocytosis involves material leaving the cell, while endocytosis deals with material entering the cell. There are three pathways of endocytosis: receptor-mediated endocytosis, bulk phase endocytosis and phagocytosis.

Receptor-mediated endocytosis involves specific receptor proteins located on the surface of a plasma membrane in depressions called coated pits. When an appropriate molecule contacts its specific receptor, it binds to the receptor. The pit sinks into the cytoplasm, forming an endocytic vesicle containing the molecule.

In bulk-phase endocytosis a small volume of fluid, regardless of content is taken into the cytoplasm as described for endocytosis above. The final type of endocytosis, phagocytosis involves the engulfing of solid materials such as microorganisms or cellular debris. This is the typical way that Amoebas some other protistans get food and that some white blood cells attack pathogens.

CHAPTER 6

GROUND RULES OF METABOLISM

1. *Define free radical. How does it relate to aging?* C1 A free radical is a highly reactive charged molecular fragment that can travel around in a cell and disrupt normal metabolic reactions or attack vital molecules such as DNA. Enzymes such as superoxide dismutase and catalase will repair damage induced by free radicals. The ability of the body to produce enzymes declines with age so that cells throughout the body may die or show indications of damage or aging. The development of dark spots on the skin are known as "age spots" and their number increases as a person ages.

2. *State the first and second laws of thermodynamics. Does life violate the second law?* 6.1 The first law of thermodynamics deals with the quantity of energy available. It states that the amount of energy in the universe is constant. Energy can neither be created nor destroyed; it can only change from one form to another. The second law of thermodynamics deals with the quality of the energy available. It may be stated as follows: The spontaneous direction of energy flow is from high-quality forms to low-quality forms. Some energy is lost when it is transferred from one form to another. Often these laws are paraphrased in the vernacular: "The first law states that we can't get something for nothing while the second law says that we can't break even (as energy is used, entropy increases)." Life does not violate the second law because the living system is continually supplied by energy from outside the system (ultimately the sun). The incoming energy prevents entropy of the living system, and living organisms use energy to maintain their high degree of organization.

3. *Define and give examples of potential and kinetic energy.* 6.1 Potential energy is stored energy, the ability to do work when released. A coiled spring, a bullet in a gun, an ATP molecule, food, a rock on top of a hill all possess potential energy that can be released when the proper trigger is activated. Kinetic energy refers to the energy of motion that can do work by imparting motion to other things. A rock held at chest level has potential energy. When the rock is released the potential energy is converted to

kinetic energy. The further it drops, the less potential energy is available and the more potential energy is converted into kinetic energy. The kinetic energy will all be released when the rock hits your toes. Muscular movement is an example of kinetic energy as is heat or thermal energy.

4. *What can a chemical reaction do to a substrate molecule?* 6.2 A substrate molecule is a specific molecule that an enzyme can recognize and bind briefly with and modify in a specific way. The chemical reaction of the substrate with an enzyme may alter its shape, split it into smaller parts or cause it to combine with another molecule.

5. *Make a simple diagram of the ATP molecule, then highlight which part of it can be transferred to another molecule and later replaced. Why is such a transfer possible at so many different sites in the cell, and what can it accomplish?* 6.3 The ATP molecule consists of three parts: adenine (a nitrogenous base), ribose (a pentose sugar) and three phosphate groups (two of which have high energy bonds). The high energy phosphate groups of ATP may combine with another molecule in a chemical reaction called phosphorylation. When the ATP molecules give up one of its high energy phosphate groups it becomes ADP (adenosine diphosphate). If the ADP molecule gives up its remaining high energy phosphate group it becomes AMP (adenosine monophosphate). If the last phosphate molecule is removed from AMP, a molecule of adenosine will be formed. Adenosine contains ribose and adenine. The ATP molecule is the common source for high energy phosphates. The ATP forms ADP during phosphorylation and the ADP will pick up a phosphate group from another source to reform ATP. The ATP/ADP cycle is the major energy source of all living cells. The energy drives hundreds of cellular activities such as active transport, muscle contraction, and synthesis and breakdown of molecules, and movement of cellular organelles.

There are hundreds of enzymes throughout the cell that may be involved in phosphorylation reactions. The high energy phosphates may be produced in mitochondria and chloroplasts, located at membranes for active transport through membranes found in the rough endoplasmic reticulum to participate in protein synthesis, associated with motile organelles such as cilia and flagella and directly or indirectly involved in almost all metabolic pathways in the cell. The availability of ATP is all pervasive throughout a cell and is found at sites through the cell. ATP is the energy currency of cells.

6. *Define activation energy, then briefly describe four ways in which enzymes may lower it.* 6.4 Activation energy is the minimum amount of energy required to initiate a chemical reaction. Two molecules must come in contact with one another before they can react. Also they must have sufficient collision energy and they must be oriented into the proper relationship to each other before they will combine. If the concentration is great enough some random collisions will meet the required conditions for two molecules to react. Increases in temperature will also increase kinetic energy of reactants and speed up chemical reactions. Enzymes serve as catalysts to speed up the rate of chemical reactions.

Enzymes reduce the amount of activation energy required for substrates to react in four ways: (1) help substrates get together by binding the substrates to enzymes thereby increasing local concentrations of the substrates; (2) the active sites of the enzymes serve to orient that substrates into favorable configurations for the substrates to unite; (3) promote acid-base reactions and encourage transfer of hydrogen ions to/from active sites to/from substrates and (4) remove water from active sites to allow closer contact between substrates and enzymes in a nonpolar environment.

7. *Define feedback inhibition as it relates to the activity of an allosteric enzyme.* 6.5 One of the ways to reduce the rate of enzymatic controlled reaction would be to reduce the amount of enzymes available to mediate the reactions. Some enzymes have another binding site in addition to the activation site. This site is called an allosteric site (allo/steric-other/structure). If a substance fits into the allosteric site, the enzyme is inactivated and the reaction is shut down. Often the substance that fits into the allosteric site is the end product of the metabolic reaction. As the reaction proceeds there is an accumulation of the product. As the amount of the product builds up some of the product will fit into the allosteric site, inactivate the enzyme and reduce the production of the end product. This process is known as feedback inhibition because the production of the product is used to turn off the enzyme that produces it.

8. *What is an oxidation-reduction reaction? Explain how such reactions proceed in an electron transport system.* 6.6 An oxidation-reduction reaction is simply a transfer of electrons. Oxidation occurs when an electron is lost and reduction occurs when an electron is gained. A simple mnemonic to help you remember this is OIL-RIG (Oxidation Is Loss-Reduction Is Gain). In the electron transfer system, a series of enzymes operate in a sequence to transfer electrons. An electron acceptor accepts an electron from a reaction and will transfer the electron to another electron acceptor in the sequence. The electron transport system allows energy to be transferred from an excited donor through an electron transport system to a final acceptor, such as oxygen in aerobic respiration. The primary electron acceptors accept electrons at higher energy levels and release them at lower energy levels. As the electron is passed along the electron transport system some energy maybe extracted in high energy phosphate bonds to form ATP.

CHAPTER 7

ENERGY-ACQUIRING PATHWAYS

1. *A cat eats a bird, which earlier speared and ate a caterpillar that had been chewing on a weed. Which of these organisms are the autotrophs? The heterotrophs?* C1 The weed is the only autotroph, the rest are heterotrophs. The caterpillar is a herbivorous heterotroph and the bird and a cat are carnivorous heterotrophs.

2. *Summarize the photosynthesis reactions as an equation. State the key events of both stages of reactions, then fill in the blanks on the diagram on the facing page.* 7.1 Summary equation: $12H_2O+6CO_2 \longrightarrow 6O_2 + C_6H_{12}O_6+6H_2O$. The key event of the light-dependent reaction is the photolysis of water in which 12 molecules of water are split into 12 atoms of oxygen which are given off as a gaseous by-product of the light-dependent reaction and 12 molecules of hydrogen. The electrons from the hydrogen are carried

by the electron transport system and the hydrogen ions buildup in the thylakoid compartment of a chloroplast to create a gradient. The hydrogen ions leave the thylakoid compartment through a protein channel called ATP synthase and produce ATP. The light-independent reactions involve carbon dioxide fixation. Ribulose-biphosphate (RuBP) reacts with carbon dioxide in the first step in a cyclic pathway to eventually produce sugar phosphate and regenerate ribulose bisphosphate.

The blanks on the diagram on the facing page preceding in a clockwise direction starting at the one o'clock position are as follows: photosystem II, light dependent reaction, photosystem I, NADPH, light independent reaction, water released, sugar phosphate (e.g. sucrose, starch, cellulose), ATP (reservoir of hydrogen ions), water is split, oxygen released.

3. *Which of the following pigments are most visible in a maple leaf in summer? Which become the most visible in autumn?* 7.3
   a. chlorophylls
   b. carotenoids
   c. anthocyanin
   d. phycobilins

In the summer the chlorophyll molecules mask all other pigments and the leaves appear green. The exact color of leaves in the fall vary as a result of differences in concentration of accessory pigments such as carotenoids and anthocyanin. The phycobilins are pigments associated with red algae and cyanobacteria.

4. *Identify which of the substances accumulates inside the thylakoid compartment of chloroplasts: glucose, chlorophyll, carotenoids, hydrogen ions, or fatty acids.* 7.5 Hydrogen ions accumulate in the thylakoid compartment and when their concentration builds up they are used in the chemosmotic production of ATP.

5. *Which is not used in the light-independent reactions? ATP, NADPH, RuBP, carotenoids, free oxygen, CO_2, or enzymes?* 7.6 The substances not required are: carotenoids and free oxygen.

6. *How many carbon atoms from CO_2 molecules must enter the Calvin-Benson cycle to produce one sugar phosphate? Why?* 7.6 Six carbon atoms from six carbon dioxide molecules enter the Calvin-Benson cycle. They will produce 12 PGAL molecules. Ten PGAL molecules are used to regenerate RuBP for the cycle, while the remaining two PGAL molecules are used to form sugar phosphate (glucose). Six carbon atoms from six carbon dioxide molecules are needed to balance the equation.

7. *A busily photosynthesizing plant takes up molecules of CO_2 that have incorporated radioactively labeled carbon atoms ($^{14}CO_2$). Identify the compound in which the labeled carbon will first appear: NADPH, PGAL, pyruvate, or PGA.* 7.6 The carbon atoms from CO_2 will combine with ribulose bisphosphate (RuBP) to form an unstable six carbon intermediate product that breaks down to form two molecules of PGA.

CHAPTER 8

ENERGY-RELEASING PATHWAYS

1. *Is this true or false: Aerobic respiration occurs in animals but not plants, which make ATP only by photosynthesis.* 8.1 The statement is false. Both plants and animals primarily derive their energy (ATP) from aerobic respiration. Plants do make ATP during photosynthesis. They and other organisms make ATP by breaking covalent bonds of carbohydrates, lipids, and proteins. Various organic compounds can be broken down into compounds that can enter various points in intermediary metabolism such as the Krebs cycle and generate ATP in the Krebs cycle or electron transport system.

2. *For this diagram of the aerobic pathway, fill in the blanks and write in the number of molecules of pyruvate, coenzymes, and end products. Also write in the net ATP produced in each stage, as well as the net ATP formed from start (glycolysis) to finish.* 8.1 The number of each designated molecule starting from the top and reading downward and left to right are as follows. 2 ATP, 2

NADH, 2 pyruvate, 2 NADH, 2CO_2, 6 NADH, 2 $FADH_2$, 2ATP, 32 ATP, 36 ATP.

3. *Is glycolysis energy requiring or energy releasing? Or do both kinds of reactions occur during glycolysis?* 8.2 2 molecules of ATP are required to initiate the process of glycolysis which releases 4 ATP molecules (a net of 2 ATP). Both kinds of reactions are involved in glycolysis.

4. *In what respect does electron transport phosphorylation differ from substrate-level phosphorylation?* 8.2, Figure 8.7   Substrate-level phosphorylation occurs in glycolysis and does not require a transport system nor require oxygen while the electron transport phosphorylation involves a series of electron acceptors ending with oxygen as the final acceptor.

5. *Sketch the double-membrane system of the mitochondrion and show where transport systems and ATP synthases are located.* 8.3 Do as instructed. Your diagram should show an inner membrane and an outer membrane. Embedded in the inner membrane are the electron transport systems and as the electrons pass through these systems from the inner compartment to the outer compartment ATP is generated and the hydrogen ions combine with oxygen to form water. Your diagram should resemble figure 8-5.

6. *Name the compound that is the entry point for the Krebs cycles and state whether it directly accepts the pyruvate from glycolysis. For each glucose molecule, how many carbon atoms enter the Krebs cycle? How many depart from it, and in what form?* 8.3
The three carbon pyruvate generated by glycolysis enters the mitochondrion where an enzyme strips it of a carbon forming carbon dioxide. The two carbon compound remaining is picked up by a coenzyme to form acetyl-CoA. The acetyl-CoA is picked up by oxaloacetic acid, the entry point into the Krebs cycle, and forms citric acid to start the cycle.

Each glucose molecule yields two three carbon pyruvate molecules during glycolysis. Two of these six carbons are released as carbon dioxide when the two pyruvate molecules form two acetyl groups. The two acetyl groups have two carbons each so that a total of four carbons are passed on to two oxaloacetate molecules that will enter the Krebs cycle. There are two turns of the Krebs cycle to use the

29

two acetyl CoA molecules. Each turn of the cycle releases two carbon dioxide molecules. Thus a total of four carbon dioxide molecules are released from the Krebs cycle for each glucose molecule used. Remember that two other carbon dioxides are released when the two molecules of pyruvate enters the Krebs cycle. A total of six carbon dioxide molecules are produced when the six-carbon glucose molecule is used.

7. *Is this statement true or false? Muscle cells cannot contract at all when deprived of oxygen. If true, explain why. If false, name the alternative(s) available to them.* 8.5 As stated, the statement is false. When a muscle lacks sufficient oxygen to allow the completion of the aerobic pathway it switches to anaerobic processes. The process is known as lactate fermentation. Glycolysis produces two pyruvate, two NADH and two ATP molecules. Pyruvate is the final acceptor of the electrons in $NADH_2$ so that it becomes the final product lactate. Sufficient energy is generated to allow the muscle to function until lactate builds-up to such a level as to produce fatigue in the muscles and contraction can no longer continue.

UNIT II PRINCIPLES OF INHERITANCE

CHAPTER 9

CELL DIVISION AND MITOSIS

1. *Define the two types of division mechanisms that operate in eukaryotic cells. Does either one divide the cytoplasm?* 9.1 The two types of nuclear division, or karyokinesis, in the eukaryotes are mitosis and meiosis. Cytokinesis is the division of the cytoplasm that may or may not follow karyokinesis. Cytokinesis does occur in most cases .

2. *Define somatic cell and germ cell.* 9.1 Somatic cells are diploid body cells, while germ cells refer to haploid cells or gametes. Meiosis occurs to specific diploid precursor cells in the gonads and results in the formation of haploid cells that will develop into the germ cells (egg or sperm).

3. *What is a chromosome called when it is in the unduplicated state? In the duplicated state? (with two sister chromatids)* 9.1 Chromosome means "colored body." They are rod-shaped bodies that become visible during mitosis because they pick up cellular stains. They are the physical structures in the nucleus that carry the genes. Prior to and after replication a chromosome is called a chromosome. As long as there is only one centromere, the whole structure, including the original DNA and its replicated strand, is called a chromosome. The old and the replicated units are called sister chromatids as long as they remain together (bound together by the same centromere). After the sister chromatids separate, both are called chromosomes.

4. *Describe the microtubular spindle and its functions, then name and briefly describe the stages of mitosis.* 9.3 A spindle consists of microtubules arranged in two sets. Each set extends from one of the two poles (end points) of the spindle. The two sets overlap each other a bit at the spindle equator, or midway between the poles. The formation of this bipolar, microtubular spindle established what will be the ultimate destinations of chromosomes during mitosis.

Prophase: centrioles divide and move to opposite sides of the nucleus, the chromosomes condense and appear shorter and thicker, the nucleolus disappears, and the spindle fibers are fabricated out of microtubulin. Metaphase: the nuclear envelope disintegrates, the spindle fibers attach to the chromosomes, and the chromosomes align themselves at the spindle equator. Anaphase: the centromeres holding the sister chromatids together separate, and the chromosomes move toward opposite poles. Telophase: the chromosomes appear at opposite poles, the chromosomes decondense and disperse, the spindle fibers disappear, and the nuclear envelope and nucleolus reappear. Mitosis is complete, having produced two identical nuclei from one parent cell that underwent mitosis.

5. *Briefly explain how cytoplasmic division differs in typical plant and animal cells.* 9.4 Cytokinesis means the division of the cytoplasm resulting in the formation of two daughter cells. Animals cells do not have a cell wall, and cytokinesis is accomplished by the development of a cleavage furrow that begins at the edge of the cell and "pinches in two." Microtubules gather at the midsection of the cells, and a small depression appears. Microfilaments pull the plasma membrane on each side of the cell towards each other, thus producing two cells. In plant cells that have to form a new cell wall, vesicles filled with wall components fuse with remnants from the spindle. The cell wall begins in the center and extends toward the cell walls until two new cells are separated by a completed cell wall.

CHAPTER 10

MEIOSIS

1. *Genetically speaking what is the key difference between sexual and asexual reproduction?* 10.1, 10.6 Sexual reproduction involves the fusion of two gametes in the process of fertilization. Asexual reproduction involves a single parent only, and the offspring are clones, or genetically identical to their single parent. The primary advantage of sexual reproduction is the variability produced by sexual recombination of genes.

2. *Suppose a diploid germ cell contains four pairs of homologous chromosomes, designated AA, BB, CC, and DD? How would the chromosomes of the gametes be designated?* 10.1 A B C D

3. *From each of his parents, actor Michael Douglas (figure 10.12a) inherited a gene that influences the chin dimple trait. One form of the gene called for a dimple and the other didn't, but one is all it takes in this case. Figure 10.12b shows what his chin might have looked like if had inherited two ordinary forms of the gene instead. What are alternative forms of the same gene called?* 10.1 Each unique molecular form of the same gene is called an allele.

4. *To the right, the diploid chromosome numbers in somatic cells of a few kinds of organisms are listed. Figure out how many chromosomes would end up in gametes of each organism.* 10.2

	BODY CELL	GAMETE
Fruit fly, *Drosophila melanogaster*	8	4
Garden pea, *Pisum sativum*	14	7
Corn, *Zea mays*	20	10
Frog, *Rana pipiens*	26	13
Earthworm, *Lumbricus terrestris*	36	18
Human, *Homo sapiens*	46	23
Chimpanzee, *Pan troglodytes*	48	24
Amoeba, *Amoeba*	50	25
Horsetail, *Equisetum*	216	108

5. *Define meiosis and then describe its main stages. In what key respects is meiosis not like mitosis?* 10.3, 10.6 Meiosis is the process of chromosome reduction that occurs during cellular reproduction, in which a diploid cell contributes one member of a homologous chromosome pair to the resulting haploid cells. Meiosis is restricted to particular parts of an organism, while mitosis occurs throughout the organism. Meiosis leads to the production of haploid gametes or spores. On the other hand, mitosis produces new diploid somatic cells and is responsible for growth. Both meiosis and mitosis are characterized by four stages: prophase, metaphase, anaphase, and telophase. Meiosis requires two divisions known as meiosis I and meiosis II. The major difference in mitosis and meiosis occurs during prophase of meiosis I. During prophase I, crossing over occurs during synapsis, resulting in a change of linkage relationships, thereby increasing variability.

6. *Take a look at all of the chromosomes in the germ cell in the diagram at the right. Is this cell at anaphase I or anaphase II?* 10.3, 10.4 The cell is at anaphase I rather than anaphase II. The chromatids are joined at the centromere rather than a single chromosome as would be the case if it were anaphase II.

7. *Outline the steps by which sperm and eggs form in animals.* 10.5 Gamete formation in animals is called spermatogenesis in male and oogenesis in females. The three key events of sexual reproduction are meiosis, formation of gametes, and fertilization. During meiosis the amount of genetic material in a diploid cell is reduced to one-half to produce a haploid cell that is the organism's contribution to the next generation. The process of meiosis determines whether the maternal or paternal chromosome is selected for each chromosome in the genome. Crossing over allows for mixing of maternal and paternal alleles for each chromosome. Gamete formation allows the haploid cell to mature to function as a sperm or an egg. Fertilization brings together new combinations of genes in the resulting offspring. This is known as sexual recombination and produces the diversity that is the hallmark of sexual reproduction.

Inside the male reproductive system a diploid cell enlarges and is
called the primary spermatocyte. It undergoes meiosis to produce
four haploid cells called spermatids. The spermatids grow,
mature, form tails, and are known as sperm. Inside the female
reproductive system, a diploid cell enlarges and becomes an
immature egg (oocyte). It has much more cytoplasm. When the
oocyte undergoes meiosis, one cell retains most of the cytoplasm,
while the other small cell consists of nuclear material and is called
the first polar body. The second meiotic division also produces a
small cell called the second polar body and a large cell that gets
most of the cytoplasm. This last haploid cell becomes the mature
female gamete, known as the egg or ovum.

CHAPTER 11

OBSERVABLE PATTERNS OF INHERITANCE

1. *Define the difference between these terms:* 11.1
   a. *gene and allele*
   b. *dominant allele and recessive allele*
   c. *homozygote and heterozygote*
   d. *genotype and phenotype*

(a) A gene is the physical basis for heredity. It is a molecule of
DNA located at a specific place (locus) on a chromosome, and it
controls a particular feature of the organisms that possess it. It is a
specific sequence of nucleotides that functions as a discrete unit. An
allele is one of the alternate forms of a gene (usually there are two,
one dominant and the other recessive). They are located at a
specific locus on a chromosome (b) The dominant allele will be
expressed and will mask a recessive allele. For a recessive allele to
be expressed it must be present in a homozygous form. The dominant
gene is usually designated by a capital letter and the recessive
allele by the same letter in lower case. (c) In homozygous forms,
both alleles in the diploid individual are alike, while the
heterozygous forms possess two different alleles for the same trait.
(d) The genotype refers to the organism's genetic constitution while
the phenotype refers to the organism's physical appearance.

2. *Define true-breeding. What is a hybrid?* 11.1 A true breeding organism is homozygous for the trait in question. For example, dwarf pea plants are homozygous for the dwarf condition, and their genes can only carry the allele for dwarfism; therefore, when dwarf pea plants are crossed they will always produce more dwarf pea plants. A hybrid is the offspring of two different true-breeding forms. For example, if a true-breeding tall pea plant (DD) were crossed with a true-breeding dwarf pea plant (dd), all the offspring would be hybrids (Dd) and would no longer breed true. The word hybrid is also used to designate the offspring produced when two different species breed as exemplified by the mule that is the hybrid from a cross of a horse and donkey.

3. *Distinguish between monohybrid and dihybrid crosses. What is a testcross, and why is it useful in genetic analysis?* 11.2, 11.3 Monohybrid crosses are crosses dealing with only one pair of alleles, while dihybrid crosses deal with the inheritance of two different genes at the same time. A testcross is used to determine the actual genotype of an organism that expresses a dominant trait. The testcross involves crossing the F_1 hybrid, or any organism that phenotypically expresses the dominant allele, with a homozygous recessive. If the offspring from the testcross express either dominant or recessive traits, the organism in question is heterozygous-dominant. If all offspring express only the dominant trait, the unknown organism had a homozygous-dominant genotype.

4. *Does segregation and independent assortment proceed during mitosis, meiosis, or both?* 11.2, 11.3 The alleles for two or more genes that are not linked (do not appear on the same chromosomes) are inherited independently of each other. In other words, the distribution of different alleles to the gametes is solely by chance. Independent assortment occurs in meiosis.

5. *What do the vertical and horizontal arrows of this diagram represent? What do the bars and curved lines represent?* 11.3 The vertical arrow refers to the number of individuals with the same value of the trait. The horizontal arrow refers to the range of values for the trait. The bars and curved lines are graphic representations of a "bell-shaped" curve of normal distribution.

CHAPTER 12

CHROMOSOMES AND HUMAN GENETICS

1. *What is a gene? What are alleles?* 12.1 A gene is the physical basis for heredity. It is a molecule of DNA located at a specific place (locus) on a chromosome and it controls a particular feature of the organism that possess it. An allele is one of the alternate forms of a gene (usually there are two, one dominant and the other recessive).

2. *Distinguish between:* 12.1, 12.2
   a. *homologous and nonhomologous chromosomes*
   b. *sex chromosomes and autosomes*
   c. *karyotype and karyotype diagram*

(a) Humans possess 23 pairs of homologous chromosomes in each cell. Homologous chromosomes are any matching pair of chromosomes. They have the same length, same structure with centromeres in the same location and have alleles for the same genes located at the same loci on the chromosome. The only difference is that one chromosome is inherited from the father and is called a paternal chromosome and the other from the mother and is called a maternal chromosome. The chromosomes may have different alleles but they posses the same genes. Nonhomologous chromosomes represent any two chromosomes that do not belong to the same homologous pair. They are different in size and structure and possess genes for different traits. (b) Sex chromosomes are the chromosomes that determine the gender of their bearer. In humans they are called X and Y. An individual that has a XX sex composition is female and a XY individual is male. An autosome is any non-sex chromosome. Humans have twenty-two pairs of autosomes plus two X chromosomes or an X and a Y. (c) A karyotype is a technique used to study the chromosomes of an individual. It involves taking a photomicrograph of the metaphase chromosomes of an individual to be used to analyze the chromosomal make-up or to compare to other individuals. A karyotype diagram is developed by cutting out the different chromosomes in the photograph and arranging the autosomes in pairs from the largest to the smallest. The sex chromosomes are placed at the end of the sequence. The karyotype diagram enables the quick inspection of the chromosome complement. If there are extra chromosomes such as three chromosome #21 the person exhibits Down syndrome. Differences from the XX or XY sex compliment result in chromosomal abnormalities as Turner's or Klinefelter's syndromes. Deletions, duplications, translocations or extra chromosomes may also be detected from a karyotype diagram.

3. *Define genetic recombination, and describe how crossing over can bring it about. Also give an example of cytological evidence that a crossover has occurred in a cell.* 12.1, 12.4 Genetic recombination refers to the different combination of genetic traits that can be produced by sexual reproduction. In asexual reproduction the offspring are genetically identical to the single parent that produced them. In sexual reproduction the process of gametogenesis produces a wide array of genetic possibilities that are combined in the offspring. When an individual is forming a gamete there is 1/2 of a chance that the 1st chromosome would be maternal, 1/2 a chance that the 2nd chromosome would be maternal and so forth. Thus when you produce the gamete that will produce your child there are over 8,000,000 possible combinations of your chromosomes. (only one of those would have all 23 maternal chromosomes). Similarly there are more than 8,000,000 possible combinations of chromosomes from your mate. Therefore, there are more than 64 trillion possible combination of chromosomes from one couple. Another factor to be considered is that crossing over occurs on each chromosome thereby changing linkage relationships. Thus the exact genetic composition of an individual offspring is essentially limitless. Even so, some of the features of the parents will appear in their offspring although it is impossible to predict what combination will appear. It is this tremendous variation resulting from recombination through sexual reproduction that gives rise to the tremendous variability in life. Such recombination allows two or more positive traits to combine in an individual and be selected for in natural selection.

All genes found on the same chromosome are said to be linked and tend to be inherited together. For example, genes found on the X chromosome are said to be sex-linked by virtue of the fact they are found on the X chromosomes. Synapsis is the pairing of homologous chromosomes that occur during prophase I. When the homologous chromosomes separate the maternal and paternal chromosomes will exchange parts thereby altering linkage relationships from their original parental status. For example, suppose the gene for hair color and freckles were found on the same chromosome. If your father had red hair and freckles and your mother was a brunette with no freckles these traits would remain linked unless crossing over separates them during gametogenesis. The closer they were located on the chromosome the less likely would they be separated. If a crossing over event occurred between these genes then some of your gametes would contain new combinations (recombination) of red hair and no freckles or dark hair and freckles. Thus it would be possible for your genetic contribution to your offspring to contain different combinations of genes than those of your parents. Thus, crossing over produces a greater variety of genetic combination resulting in sexual recombination.

When chromosomes pair during synapsis they appear twisted around each other. When they separate a visible cross may be seen where crossing over occurs. The text describes research in 1931 in corn in which a cytological marker confirmed the phenomenon of crossing over in chromosomes 9 of corn plants. One of the chromosomes was normal, but the other was longer by virtue of an earlier translocation (a fragment of another chromosomes was attached to chromosome 9). Normally, the long chromosome was linked to certain alleles. If crossing over occurred between these alleles the long portion of the chromosome would also be transferred. Thus anytime there was a new combination the longer chromosome would confirm that crossing over occurred.

4. *Define pedigree. Also explain why a genetic abnormality is not the same as a genetic disorder or genetic disease.* 12.6 Pedigree refers to the genetic history of an individual. In genetics, pedigree refers to a diagram of the occurrence of a particular trait in a family history. A genetic abnormality is an unusual variety of a certain trait such as a blaze, a white forelock in normally colored hair. On the other hand, a genetic disorder is an inherited condition that will sooner or later cause mild to severe medical problems. Some genetic disorders are phenylketonuria, sickle-cell anemia, cystic fibrosis, hemophilia and albinism. A genetic disease is an inherited genetic condition that reduces an individuals ability to fight of an infectious disease. An example would be lupus.

5. *Contrast a typical pattern of autosomal recessive inheritance with that of autosomal dominant inheritance.* 12.7 An autosomal recessive trait may skip generations, may appear in children of two normal parents, may be inherited but not expressed because of the masking effect of a dominant gene. In an autosomal dominant trait one of the parents of the individual that expresses the trait must also express the trait. A heterozygote will express the trait and it will appear in each generation. One half of all the children of an individual with the gene will inherit the trait and express it. If neither parent express the trait there is no chance for their children to inherit the trait. Autosomal dominant traits are much rarer than autosomal recessive traits. Some examples of autosomal dominant traits include Huntington's disorder, progeria, chrondrioplasic displasia, Tay Sachs disorder and familiar hypercholesterolemia. Some autosomal recessive traits include phenylketonuria, hemophilia, cystic fibrosis, albinism, sickle cell anemia, and galactosemia.

6. *Contrast a typical pattern of X-linked recessive inheritance with that of X-linked dominant inheritance.* 12.8 Let us look at the frequency of expression of dominant and recessive genes. If the dominant allele C had a frequency of 30% the recessive allele c would have a frequency of 70%. The phenotypes for each of the genotypes would be as follows:

CY   30% of the males would express the dominant trait

cY   70% of the males would express the recessive trait

CC   9% of the females would be homozygous dominant
       (30%) x (30%)

Cc   42% of the females would be heterozygous 2 (30%) x (70%)
       51% of the females would express the dominant trait
       (9%+42%)

cc   49% of the females would express the recessive trait
       (70%) x (70%)

If the recessive allele d has a frequency of 30% the dominant allele D would have a frequency of 70%. The phenotypes for each of the genotypes would be as follows:

DY	70% of the males would express the dominant trait
dY	30% of the males would express the recessive trait
DD	49% of the females would be homozygous dominant (70%) x (70%)
Dd	42% of the females would be heterozygous 2 (70%) (30%) 91% of the females would express the dominant trait (42%+49%)
dd	9% of the females would express the recessive trait.

In a dominant sex linked trait the females would be more likely to express the trait than males because they have two chances to inherit the sex linked trait (XX) as opposed to one for males (XY). Males inherit their sex-linked traits from their mother and will express it. If a father has a sex-linked trait he will pass it on to all of his daughters and none of his sons. A recessive trait can be masked by a dominant allele in females.

7. *Define aneuploidy and polyploidy. Make a sketch of an example of nondisjunction.* 12.9 A set of chromosomes is designated by the letter N. In humans, the letter N refers to a set of 23 chromosomes. If the number of chromosomes in a cell is divisible by 23 it is said to be a euploid cell. If there more or less than complete sets of chromosomes in a cell, it is described as aneuploid. For example, normal euploid human cells have two sets of chromosomes for a total of 46 chromosomes. If a cell had either 45 or 47 it would be aneuploid. A haploid cell has N chromosomes, a diploid cell has 2N chromosomes, a triploid cell has 3N chromosomes, a tetraploid cell has 4N chromosomes. Triploid and tetraploid cells are described as polyploid. Any euploid cell with more than two sets of chromosomes is polyploid.

To illustrate nondisjunction show that two paired homologous chromosomes fail to separate during anaphase so that one pole gets an extra chromosomes while the other pole gets one less chromosome. The two cells resulting will exhibit aneuploidy.

8. *Distinguish among a chromosomal deletion, duplication, inversion, and translocation.* 12.10

Normal chromosome:	ABCDE.FG
Chromosome with deletion:	ABC.FG
Chromosome with a duplication:	ABCDBCDE.FG
Chromosome with a translocation	ABCDE.FRS
	GTU.VW

The period in the representations of chromosomes denotes the position of the centromere.

In a chromosome deletion, the chromosome breaks and the fragment is lost. In a chromosome duplication, a set of duplicate genes are added to the chromosome requiring a break in two homologous chromosomes. In a translocation a chromosome fragment from a nonhomologous chromosome becomes incorporated in the chromosome. In addition to the addition or subtraction of genetic material these abnormalities make it difficult for chromosomes to synapse (pair) properly during meiosis resulting in decrease in viability of gametes and reduction in reproductive potential.

CHAPTER 13

DNA STRUCTURE AND FUNCTION

1. *Name the three molecular parts of a nucleotide in DNA. Name the four different bases that occur in these nucleotides of DNA.* 13.2 The three parts of a nucleotide are a pentose sugar (deoxyribose or ribose), a phosphate group, and a nitrogenous base. The nitrogenous bases in DNA are adenine, cytosine, guanine, and thymine (in RNA, uracil is substituted for thymine).

2. *What kind of bond joins two DNA strands in a double helix? Which nucleotide base pairs with adenine? with guanine?* 13.2, Hydrogen bonds between the nitrogenous bases bind the two strands of DNA together. Thymine pairs with adenine. Cytosine pairs with guanine.

3. *Explain how DNA molecules can show constancy and variation from one species to the next.* 13.2 All DNA molecules exhibit the same bonding patterns. The DNA molecules that pass from parent to offspring are identical barring the occasional mutation. This constancy in the inheritance of the genetic message accounts for the constancy in a species. Species differ from each because they are sexually isolated and the mutations that occur in a population will accumulate and eventually produce enough variation between populations to produce new species.

CHAPTER 14

FROM DNA TO PROTEINS

1. *Are the polypeptide chains of proteins assembled on DNA? If so, state how. If not, tell where they are assembled, and on which molecules.* C1, 14.2 The central dogma describes how the gene functions to produce proteins. The DNA molecule has the genetic code inscribed in its sequence of nucleotides. The process of transcription converts the code in the DNA molecule into a complementary RNA molecule called messenger RNA. The basic sequential information is retained in this exacting process. The molecule of messenger RNA leaves the confines of the nucleus to travel to a ribosome located in the cytoplasm or rough endoplasmic reticulum where protein synthesis occurs with the assistance of transfer RNA. The transfer RNA brings the amino acids to the ribosome, where ribosomal RNA assembles the amino acids into the sequence dictated by the sequence of nucleotides on the messenger RNA.

2. *Define gene transcription and translation, the two stages of events by which proteins are synthesized. Both stages proceed in the cytoplasm of prokaryotic cells. Where does each stage proceed in eukaryotic cells?* C1 Transcription is the formation of a messenger RNA molecule from the sense strand of DNA. The base sequence in the RNA is complementary to the DNA molecule. Transcription occurs in the nucleus of eukaryotic cells. Translation is

the conversion of the message contained in the messenger RNA molecule to a particular sequence of amino acids in a polypeptide sequence. This process occurs in the ribosomes in the cytoplasm of eukaryotic cells.

3. *Briefly state how gel electrophoresis, a common laboratory procedure, works. How did it yield a clue that small differences in normal and abnormal versions of the same protein may lead to big differences in how the proteins function?* 14.1 In gel electrophoresis a molecule is placed in a viscous gel and subjected to an electrical field. Differences in molecular size, shape and surface charges control the amount of migration that the test molecule will experience. The degree of migration may be used to separate two different molecules. Linus Pauling and Harvey Itano used this technique to demonstrate the molecular difference between normal hemoglobin and sickle-cell hemoglobin. The difference between the two hemoglobins is found in the sixth amino acid of the beta chain. In normal hemoglobin the amino acid is glutamate which carries a net negative charge. In sickle-cell hemoglobin, a mutation produces a new amino acid, valine in the sixth position. Valine has no net charge so that sickle-cell hemoglobin behaves differently in an electrophoretic field. In the blood capillaries where oxygen concentrations are lowest the lack of charge contributed by valine renders the molecule hydrophobic and sticky. The result of the low oxygen concentration causes the aberrant hemoglobin to distort the red blood cells producing the symptoms that occur during a sickling-cell crisis.

4. *Name the three classes of RNA and briefly describe their functions.* 14.2, 14.3 The three RNA molecules involved in protein synthesis are messenger RNA, ribosomal RNA, and transfer RNA. During the process of transcription the sequence of nucleotides in the DNA code is maintained in messenger RNA. The ribosomal RNA molecules are found in the ribosome and control the assembly of proteins during translation. The transfer RNA molecules have specific amino acids attached to them and deliver the amino acid to the ribosome where they are assembled according to the sequence contained in the mRNA molecule.

5. *In what key respects does the sequence of nucleotide bases in RNA differ from those in DNA?* 14.2 The sequence of nucleotides in RNA are complimentary to the sequences of nucleotides in DNA.

44

Thus, if the sequence in DNA is adenine-guanine-cytosine-thymine-adenine-cytosine, the sequence in RNA would be uracil-cytosine-guanine-adenine-uracil-guanine. Remember that RNA's uracil is complimentary to adenine in DNA.

The major difference between DNA and RNA is that DNA is a double strand of nucleotides while RNA consists of a single strand. In addition there are two other different features. The pentose sugar in the RNA nucleotide is ribose, while in DNA it is deoxyribose. In RNA, the nitrogenous base uracil is substituted everywhere there is a thymine nucleotide in the DNA molecule.

6. *How does the process of gene transcription resemble DNA replication? How does it differ?* 14.2 In gene transcription, the DNA molecule separates and unwinds exposing the sense strand of DNA for transcription. Under the auspices of RNA polymerase, complementary RNA nucleotides are lined-up one nucleotide at a time in the 5 ' to the 3 ' direction. In DNA replication, the enzyme DNA polymerase will assemble one-half of the new DNA molecule by adding DNA nucleotides as above, one nucleotide at a time. At the same time, the other strand will be assembled using Okazaki fragments and be assembled in the 3 ' to 5 ' direction. In addition, in transcription, only a portion of the DNA molecule will be transcribed, while in DNA replication the entire DNA molecule is replicated. Only one RNA strand is produced by transcription whereas two double-stranded DNA molecules are produced by replication.

7. *The pre-mRNA transcripts of eukaryotic cells contains introns and exons. Are the introns or exons snipped out before the transcript leaves the nucleus?* 14.2 In addition to attaching a cap on the 5 ' end of the RNA molecule and a tail on the 3 ' end, the maturation of a mature mRNA molecule requires the removal of the introns (nontranscribed sequences of nucleotides) and the joining of the exons to form a mature mRNA molecule that leaves the nucleus and goes to a ribosomes where it is transcribed into a protein.

8. *Distinguish between codon and anticodon.* 14.3 A codon is a sequence of three nucleotides found in the mRNA molecule. They are specified by the appropriate three nucleotides in a DNA molecule. The three nucleotides in mRNA are complimentary to three nucleotides in an anticodon in transfer RNA. There are 64

possible combination of the four nitrogenous bases in a triplet code. Sixty-one of the codons will produce a complimentary anticodon. The anticodons of transfer RNA specify different amino acids. Three of the codons function as stop signals for translation and the other codon (AUG) serves to specify either the amino acid methionine or a start signal for translation.

9. *Cells use the set of sixty-four codons in the genetic code to build polypeptide chains from twenty kinds of amino acids. Do different codons specify the same amino acid? If so, in what respect do they differ?* 14.3 The last of the three nucleotides in a codon is the least sensitive of the nucleotides, that is, if the third nucleotide changes, the amino acid specified may remain the same. The change of the first nucleotide occasionally will specify the same amino acid. The middle nucleotide is the most critical, and anytime that it is changed the amino acid specified will also change. Because there is more than one codon for each amino acid the code is said to be redundant.

10. *Name the three stages of translation and briefly describe the key events of each one.* 14.4 The steps of translation occurs in three stages, which are called initiation, elongation, and termination. mRNA is transcribed by template DNA. It is complementary to one strand of DNA and contains the genetic instructions in the DNA code. rRNA is transcribed by the DNA of the nucleolus and functions in the ribosome to translate the message in the mRNA into a sequence of amino acids that forms a protein. It functions by bringing the codon of mRNA together with the anticodon of tRNA in a way to line up amino acids for protein synthesis. tRNA is called transfer RNA; it has a specific anticodon and transports a specific amino acid to the site of protein synthesis in the ribosome.

In initiation, a tRNA that can start translation and a mRNA transcript are loaded into a ribosome. The initiator tRNA binds with the small ribosomal subunit. The mRNA with the initiator signal, AUG, binds with both small and large subunits to form an initiator complex. This completes initiation and elongation may proceed. In elongation, mRNA moves between the subunits of the ribosomes. Its codons are matched by complimentary tRNA anticodons carrying the appropriate amino acids. Enzymes catalyze the formation of peptide bonds between the amino acids in the growing peptide chain. In termination, when a stop codon is reached there is no complimentary anticodon. Enzymes will release the RNA molecule and the polypeptide chain from the ribosome and transcription is terminated.

11. *Fill in the blanks on the diagram below.* 14.19 Transcript processing from left to right: mRNA, tRNA, mature RNA, ribosomal subunits and mature tRNA. Translation from left to right: Convergences of cytoplasmic pools of amino acids, tRNAs, ribosomal subunits and protein.

13. *Define and state the possible outcomes of the following types of mutation: a base-pair substitution, a base insertion, and an insertion of a transposable element at a new location in the DNA.* 14.5 One type of mutation consists of the substitution of one nucleotide for another. If the new nucleotide changes, the amino acid specified by the codon in which the nucleotide is found may change the amino acid specified. Then the protein produced by the codon would differ in its primary structure. If this amino acid substitution is in a critical part of a protein, it may change its function. A second type of mutation known as frameshift mutation occurs if a nucleotide is added or subtracted from the nucleotide sequence. Every codon downstream from the change will have altered codons and therefore a change in most of the amino acids downstream of the addition/subtraction. It is likely that massive changes in a protein resulting from a frameshift mutation will so alter the proteins so that it will no longer function normally.

The movement of a transposable element may insert this DNA element anywhere in the same DNA molecule or another one. Often they will inactivate the DNA molecule into which they have been inserted.

14. *Define and explain the difference between mutation rate and mutation frequency. What determines whether an altered product of a mutation will have helpful, neutral, or harmful effects?* 14.5 Every gene has a characteristic mutation rate. It may be the location or the environment of the gene on a chromosome or it may be the inherited tendency toward the production of mutations (i.e. some individuals have a higher frequency of certain kinds of cancer). The mutation rate refers to the probability that a specific gene will mutate spontaneously during a specified interval, such as each DNA replication cycle. On the other hand, mutation

frequency refers to the number of times a gene mutation has occurred in a given population.

Two things determine whether the mutation will be beneficial or detrimental: (1) the amount of change, and (2) the protein involved. If the mutation occurs in a gene controlling the vital enzyme and the enzyme is unable to synthesize its product the result of the mutation, may be the death of the individual. If, the mutation occurs in a gene controlling the production of insulin it could result in diabetes if it occurred in a cell in the pancreas. If, on the other hand, this mutation occurred in a liver or brain cell it would not have a chance to be expressed. The most critical mutations are those that occur in germ cells so that all of the cells of the offspring would bear the mutation and have the chance to express it.

CHAPTER 15

CONTROLS OVER GENES

1. *In what fundamental way do negative and positive control of transcription differ? Is the effect of one or the other form of control (or both) reversible?* 15.1 In a negative control system a regulatory protein may bind to a certain site on the DNA molecule and block its transcription. On the other hand, in a positive control system a regulatory protein may bind to a site on the DNA molecule and promote the initiation of transcription. Both systems can be reversed when the conditions that called for the activity change.

2. *Distinguish between:* 15.2
   a. *promoter, and operator,*
   b. *repressor protein and activator protein.*

The promoter is a specific sequence of nucleotides to which RNA polymerase attaches to initiate transcription. An operator is a specific sequence of nucleotides that functions to induce or repress structural genes according to the Jacob-Monad hypothesis. A repressor protein binds to part of the promoter region to prevent transcription by a structural gene. Bound activator proteins promote transcription by assisting RNA polymerase in binding to the structural gene and starting transcription.

3. *Describe one type of control over transcription of the lactose operon in* E. coli, *a prokaryotic cell.* 15.2 The Jacob-Monad hypothesis explains methods of control found in *E. coli.* When *E. coli* is grown in an environment containing lactose, it produces the enzyme lactase, which breaks the lactose substrate down to form the products glucose and galactose. The system that controls the production of the enzyme lactase is called an operon. The operon consists of a promoter, the operator system, and three structural genes. A regulator gene, located elsewhere, codes for a repressor protein that will inhibit the lactose operon by binding on it when the lactose concentration is low (no need for lactase at this time). The repressor overlaps the promoter and blocks the access of RNA polymerase to the gene. When lactose is present, it binds to the repressor protein, altering the shape of the repressor protein so that it no longer interferes with the genes and so that they can transcribe the mRNA needed to manufacture the lactase enzyme.

4. *A plant, fungus, or animal is composed of diverse cell types. How might this diversity arise, given that the body cells in each organism inherit the same set of genetic instructions? As part of your answer define cell differentiation and explain how selective gene expression brings it about.* 15.3 Various studies have confirmed that cells taken from different parts or organs of an organism have the same genes. The differences that develop in the cells throughout an organism are not due to differences in the genes found in those cells. The differences are controlled by which genes are induced (turned on) or repressed (turned off). Other control elements include enzyme systems, regulator proteins, accumulation of products, rate of DNA transcription and translation, environmental differences, chemical gradients, and the activities of adjacent cells. Differentiation results in changes in cells so that they take on different functions, appearances, compositions, and even positions. It arises through selective gene expression.

5. *Review figure 15.4. Using the diagram below, define five types of gene controls in eukaryotic cells and indicate where they take effect.* 15.3 Some controls are related to transcription. Gene amplification produces multiple copies of DNA molecules that code for a vitally needed product. The availability of multiple copies of the gene enables certain cells to produce large amounts of the needed product. DNA segments may be rearranged as is characteristic of white blood cells that produce a wide variety of antibodies. DNA molecules may be modified through interaction with histones and other proteins that may isolate the DNA and keep it from reacting. Pre mRNA molecular transcripts may be modified before they leave the nucleus. The splicing of exons of the mRNA can be done in several alternate arrangements. Different cells may end up with alternative splices of mRNA resulting in subtle differences in their proteins and therefore differences in activities of different cells throughout the body. After translation the transcripts may be inactivated or its stability modified. Future conditions in the cell may reactivate these inactivated products and allow rapid metabolic use of these stored transcripts. Other polypeptide transcripts become modified as they pass through the cytomembrane system. Phosphate groups or oligosaccharides may be added to them.

The five kinds of controls in eukaryotes are as follows:
(1) transcription controls, (2) transcript processing controls,
(3) transport controls, (4) translational controls and (5) post-translational controls.

6. *If a polytene chromosome and a lampbrush chromosome are both evidence of transcriptional activity, in what respect do they differ?* 15.4, 15.5 Polytene chromosomes are found in the salivary glands of members of the Diptera, an order of flies such as midges. The larvae of midges require large amounts of saliva and the DNA in their salivary glands has been replicated repeatedly. The copies of the DNA molecules remain together in parallel arrays creating a large chromosome known as a polytene chromosome. Lampbrush chromosomes are found in the unfertilized eggs of amphibians. These eggs grow very fast and some genes actively transcribe large amounts of RNA and ribosomes which function in protein synthesis. In order to allow for rapid transcription, the nucleosomes (basic unit or organization of eukaryotic chromosomes) become decondensed and somewhat dissociated from the histone

and other proteins. The spreading out of the DNA loops to allow massive transcription produce a bristly, lampbrush configuration. The need for rapid transcription is responsible for both polytene and lampbrush chromosomes. The difference between them is both in their appearance and the fact that the polytene chromosome involves massive replication of DNA molecules while the lampbrush chromosome exhibits a method to make the DNA in a chromosome available for rapid transcription without having to synthesize more DNA.

7. *What is a Barr body? Does it appear in the cells of males, females, or both? Explain your answer.* 15.4 A Barr body is a structure found in nuclei that represents a condensed inactive X chromosome. Since normal males have only one X chromosome that has vital genes that must function for survival a normal male would not have a Barr body. Normal females are XX and one of their X chromosomes is active while the other forms a Barr body. It would be possible to determine the sex of normal males and females based upon the presence of a Barr body in the nuclei of their cells. Those individuals with other chromosome compliments may have different conditions. An individual with Turner's syndrome (XO) would not have a Barr body. An individual with Klinefelter's syndrome (XXY) would have a Barr body. A metafemale (XXX) would have two Barr bodies.

8. *Define hormones. Why do hungry midge larvae depend on the hormone ecdysone?* 15.5 A hormone is signaling molecule produced by endocrine glands or cells that are distributed throughout the body of an organism. When the hormone reaches its specific target tissue, the cells with the appropriate receptors respond in an appropriate fashion. Ecdysone is the insect hormone that controls molting. When ecdysone binds to receptor cells in the salivary gland, it triggers rapid replications of multiple copies of DNA in the salivary gland. This region of DNA in the chromosome puffs out and transcribes massive amounts of mRNA that code for proteins found in saliva. Midge requires massive amounts of saliva.

9. *What are the characteristics of cancer cells? Explain the difference between a benign tumor and one that is malignant.* 15.6 Cancer could be defined as uncontrolled growth. Some of the characteristics of cancer include the following: (1) abnormal membrane transport and permeability and changes in surface

characteristics of cancer cells; (2) cytoplasmic changes including cytoskeletal shrinkage and metabolic changes such as greater amount of glycolysis; (3) rapid cell division with a turnoff of the inhibitors for overcrowding as well as proliferation of capillaries to supply the tissue; and (4) failure of tissue to adhere to the parent material. Obviously cancer cells divide more rapidly than normal cells. The big problem occurs when the cancerous cells break out of their close-knit packet and spread to and invade other tissue in a process called metastasis. Benign tumors are not life threatening and do not occur in vital places, while malignant tumors can be lethal.

CHAPTER 16

RECOMBINANT DNA AND GENETIC ENGINEERING

1. *Distinguish these terms from one another:*
   a. *recombinant DNA technology and genetic engineering* 16.1
   b. *restriction enzyme and DNA ligase* 16.1
   c. *cloned DNA and cDNA* 16.2, 16.4
   d. *PCR and DNA sequencing* 16.2

(a) In recombinant DNA technology, DNA from different species is cut (using restriction enzymes) and spliced (using ligase) to the DNA of the receiving species. The newly combined DNA is then amplified via copying mechanisms to produce adequate supplies of the DNA. Genetic engineering refers to the deliberate modification of genes followed by their insertion into the same organism or another one. (b) Restriction enzymes are used to cut DNA molecules at specific sites having the appropriate specific short nucleotide sequences. Ligases are enzymes that splice or tie two DNA fragments together at the cut sites. (c) Cloned DNA molecules are produced in one of two ways. The specific DNA desired is selected from a DNA library and amplified by a living "factory"- bacteria or yeast that can reproduce rapidly and take up plasmids. The second way involves the polymerase chain reaction using test tubes not microbes to amplify tiny amounts of DNA into large samples. On the other hand, cDNA is produced by using mature messenger RNA molecules. An enzyme, reverse transcriptase, generates a matching DNA strand that is assembled onto the mRNA. Other enzymes then remove the RNA from the hybrid RNA-DNA molecule and then synthesize a complementary DNA strand to make a complete double stranded DNA molecule. (d) PCR stands for polymerase chain reaction, the new method for amplifying genes. A double stranded DNA molecule is unwound by being heated. The separate single strands serve as templates for synthesized short nucleotide sequences that pair with them. These short sequences serve as primers for replication. The process is repeated over and over again until enough double stranded DNA is available. DNA sequencing refers to the determination of nucleotide sequence of different length fragments of DNA produced by exposing DNA to restriction enzymes that cut the DNA at specific nucleotide sequences. The different length fragments produced by the action of the restriction enzymes are placed in elelectrophoresis and separated according to length. The Sanger method of nucleotide sequencing involves DNA polymerase, short nucleotide sequences that serve as primers for replication and each of the four nucleotide subunits of DNA (e.g. A, T, C, and G). Radioactivity labeled bases identify which base is paired with on the template strand when the complementary strand is constructed. The sequence of bases can be determined by reading the bands in the electrophoresis gel. The process is described in detail in figure 16.5.

2. *Define RFLPs. briefly explain one method by which they are produced, then name some applications for RFLP analysis.* 16.3
RFLPs stands for restriction fragment length polymorphisms. A large sample of an individual's DNA molecule are cut by restriction enzymes and subjected to gel electrophoresis to separate the fragments by size (smaller fragments move faster in the electrical field of electrophoresis). The fragments are placed in solution and reacted with a labeled probe. The DNA fragments may be identified by length and banding patterns and are unique for any individual. They are often referred to as a DNA fingerprint.

The RFLPs are used in the human genome project which is in the process of determining the genetic sequence for the human genome. The potential exists to locate specific genes such as those producing cystic fibrosis, sickle-cell anemia, phenylketonuria and other genetic disorders. Prenatal diagnosis will become more precise. Evolutionists can compare present day genomes with genetic samples that have been preserved thousands of years ago to gain insight on the evolutionary process. Paternity cases may be resolved through this technique. Perhaps the most publicized application is in forensic analysis; the use of blood, saliva or semen samples to demonstrate a specific individual was at the scene of a crime.

3. *Explain how DNA probes and nucleic acid hybridization can be used to identify genetically modified host cells.* 16.4 DNA probes are short DNA sequences synthesized from radioactivity labeled nucleotides. Part of a probe is designed to pair with some portion of the DNA molecule of interest. The base pairing between nucleotide sequences from different sources (i.e. DNA probe and DNA molecule of interest) is called nucleic acid hybridization.

Recombinant DNA molecules are exposed to bacteria or other living cells that incorporate them. Plasmids contain a gene that confers resistance to an antibiotic. When a culture of cells is growing in an environment that contains the antibiotic, all cells without the plasmid containing the antibiotic-resistant gene die and only those cells with the antibiotic-resistant gene survive. These cells reproduce and form colonies (clones). Cells are collected from the colonies, exposed to solutions that rupture the cells and replace their DNA. The DNA is treated so that it unwinds and the DNA probes are added, which then hybridize with the section of DNA with the complimentary base sequence. The radioactivity enables the researcher to locate the colonies that contain the DNA of interest.

UNIT III PRINCIPLES OF EVOLUTION

CHAPTER 17

EMERGENCE OF EVOLUTIONARY THOUGHT

1. *Define and contrast the theories of catastrophism and of uniformity.* C1, 17.1 The idea of catastrophism is based on the idea that a great number of catastrophes occurred during the relatively short span that humans lived on earth. Extrapolating from the human experience it was proposed that all the changes in the earth's surface must have been the result of sudden, violent geological processes (volcanic eruptions and earthquakes). The theory of uniformity is based on the concept that geological processes proceed at a uniform rate. Using the current rates of erosion and volcanic activity geologists indicate that the earth is much older than the 6,000-10,000 years based upon biblical history. The expanded time frame provided by geologists allowed evolutionists to propose a gradual time scale for evolution.

2. *Define biogeography and comparative anatomy. How did studies in both disciplines contradict the idea that species have remained unchanged since the time of creation?* 17.1 Biogeography is the study of the distribution of living organisms throughout the world. Biogeography indicates that organisms often exhibit a different distribution pattern now than in the past as revealed by fossil records. For example, shark's teeth may be found in the Coastal Plain two hundred miles from the ocean. Horses, rhinoceroses, and camels evolved in North America but are now found elsewhere. Ninety-five percent of the flora and fauna of Hawaii are endemic to the islands, but there is nothing unusual about the islands except their isolation. Darwin's finches on the Galapagos Islands are excellent examples of invasion of a remote area followed by changes leading to speciation. There are many albino cave salamanders, fish, and insects. These animals more

closely resemble forms of life that live outside the cave but in the same geographic region. They do not resemble each other except for their lack of sight and pigmentation. There does not appear to be a uniform species of cave animals. The conclusion to be drawn is that local species became entrapped in the cave environment and through time evolved the albinism and blindness characteristic of many cave animals. The basic question is: If all species were created at the same time and same place, why are there so many types today with such different patterns of distribution?

Comparative anatomy is the systematic study of the similarities and differences in the structure of animals. In comparative anatomy, homologous structures indicate that certain animals share a common ancestor. Vestigial structures indicate that some snakes had legs and that whales had functional pelvic girdles in past ages. It would be appropriate to compare present day forms to their fossil ancestors. The fossil sequence in horses shows an increase in size, loss of toes, change in dental pattern to allow switching from browsing to grazing, elongation of the muzzle, and the development of a gap between the incisors and molars.

3. *Cuvier and Lamarck interpreted the fossil record differently. Briefly state how their interpretations differed.* 17.2 Cuvier proposed the theory of catastrophism, which stated that worldwide catastrophes caused many species to become extinct. In time, the survivors were able to repopulate the world. Then another catastrophe eliminated more species and was again followed by repopulation of the world by survivors. A series of catastrophes, each followed by repopulations, led to the present surviving forms of life on earth.

Lamarck proposed that life was created long ago in a simple state. A built-in drive for "perfection," located in nerve fibers, gradually changed the simple forms into more complex forms.

4. *Define evolution. Define evolution by natural selection. Can an individual evolve?* 17.1, 17.3 Evolution may be defined as descent with modification. Another definition states that through time there is a change in gene frequency in a population. Evolution by natural selection involves differential survival and reproduction of organisms that possess genes that are adaptive to specific environments. Individuals are not able to evolve—they may only

56

make temporary adjustments to their environment. The biological unit that undergoes evolution is the population. Two isolated populations may accumulate enough differences so they are no longer able to interbreed. When this happens, speciation has occurred.

CHAPTER 18

MICROEVOLUTION

1. *Define genetic equilibrium. Explain how these processes can send allele frequencies out of equilibrium.* 18.1, 18.2
   a. *mutation*
   b. *natural selection*
   c. *gene flow*
   d. *genetic drift*

Genetic equilibrium refers to stability of allelic and genotypic ratios in a population over succeeding generations. (a) A mutation will change the frequency of alleles in a population. For example, if there is a new mutation in a population to produce a recessive allele for albinism. This allele will be added to the normal alleles for color. Eventually, through chance the allele may occur in the homozygous condition and be expressed. (b) In natural selection a population's genetic frequency is controlled by external environmental factors. If for example, a new pesticide were developed and released in the environment, eventually over time a population would by chance develop mutant forms capable of surviving exposure to the pesticide. Through natural selection, the normal members of the population would be killed by exposure to the pesticides. The mutant forms would have no adverse effects and be selected for by reproducing and contributing their genes (including pesticide resistance) to the next generation. Over time, the frequency of the mutant gene would build-up in the population. (c) Gene flow refers to the flow of genes from one population to another by migration and interbreeding. A simple example is the spread of killer bees throughout South America and into North America after their escape in Brazil. The African bees bred with native bee populations and the offspring retained many of the aggressive features of the African bees. (d) Genetic drift occurs in small populations. Chance determines which genes are passed on and which are not passed on to the next generation. Thus, there may be changes in gene frequencies in small populations simply by chance.

2. *Define fitness, as biologists use the term.* 18.1 The determination of the relative fitness (advantage or disadvantage) of a phenotype depends upon the environment in which the organism is found. The ultimate test of the success of any given phenotype is measured in its contribution of genes to subsequent generations. Advantage or disadvantage is determined not only by the ability to survive, but also by mating selection and the number of offspring produced..

3. *Define bottleneck and the founder effect. Are these cases of genetic drift, or do they set the stage for it?* 18.8 The bottleneck effect describes what happens to genetic diversity when a large widely distributed species is reduced to a small population with limited distribution. Siberian tigers or California condors are examples of species that have undergone dramatic reduction in numbers of organisms. As the numbers are reduced the genetic reservoir decreases and the genetic variation is reduced. In the founder effect, a small population with a limited gene pool is introduced to an area (such as a single pregnant female). Both the bottleneck and founder effect lead to small populations with limited gene pools that set the stage for genetic drift. In these small populations random events can eliminate certain alleles or chance occurrences increase the frequencies of other alleles. These variations in allelic frequencies that are due to chance occur in small populations and may result in dramatic changes in the population over time.

4. *Consider the brilliantly hued male sugarbird and the subdued-hued female (Figure 18.16). In the evolutionary view, what is maintaining the morphological difference over time.* 18.6 For any characteristic to be favored in natural selection it has to be better adapted to the environment than opposing characteristics. The female's primary duty in some bird species is to lay eggs and roost on them until they hatch. It would be desirable for the female to be dull in color to escape detection so that she can complete the task of

hatching the offspring without interruption. On the other hand, the male has to convince a female bird to accept him on the basis of the territory he possesses and defends as well as his physical attributes involving song and appearance. Therefore, it is advantageous for the male to be brightly colored to warn off other males attempting to invade his territory and to win selection by a female so his genes will pass to the next generation. This sexual dimorphism is an essential adaptation of bird species and has selective value that leads to its maintenance through time.

5. *As long-term studies indicate, that prospects for human newborns of very high or very low birthweight are not good. Being born too small increases the risk of stillbirth and early infant death (Figure 18.17). Similarly, pre-term rather than full-term pregnancies also increase the risks. Which of the microevolutionary process is at work here?* 18.5 This is an example of stabilizing selection in which both extremes have a selective disadvantage and the intermediate forms are favored by selection.

6. *In Figure 18.18, identify the modes of selection (stabilizing, directional, and disruptive) in each diagram).* 18.4, 18.5 The diagram to the left represents directional selection, the one in the middle represents disruptional selection, and the one to the right represents stabilizing selection.

CHAPTER 19

SPECIATION

1. *Describe how the biological species concept differs from a definition of species based on morphological traits alone.* 19.1 Ernst Mayr defined the species concept as follows: "Species are groups of interbreeding natural populations that are reproductively isolated from other such groups." The major advantage of this concept is that it is not dependent on phenotypic characteristics; rather, it emphasizes the requirement for members of the same species to interbreed. There are problems in applying this concept to species that do not reproduce sexually. It would also be hard to apply to

some species that do not breed in captivity. It is also not applicable to fossils or preserved specimens.

2. *Define speciation. Define and give examples of reproductive isolating mechanisms. and cite an example of each of the three speciation models.* C1, 19.2 Speciation is the evolutionary process that results in the formation of two species from an already existing species. It involves divergence of two reproductively isolated populations that accumulate sufficient genetic differences to prevent them from interbreeding. Reproductive isolation may be classified as prezygotic or postzygotic isolation. As the name implies prezygote isolation involves processes that occur prior to or during fertilization while postzygotic isolation occurs after a zygote is formed. Some of the mechanisms of isolation are as follows: (1) geographic isolation-two populations are separated by some physical barrier, (2) temporal isolation -two populations are sexually active at different times, (3) behavioral isolation (perhaps the most significant type of isolation)-the two populations either fail to recognize or respond to sexual behavior or signals including pheromones, (4) mechanical isolation-a physical difference between two populations prevents reproduction, (5) ecological isolation-the two populations inhabit different microenvironments in the same habitat. In gametic mortality the gametes of the two populations have developed molecular incompatibility (i.e. the tissue of the stigma and style may inhibit the growth of a pollen tube to the egg). After fertilization post zygotic isolation may result if there are problems in the development of the embryo or lack of fitness so that the hybrid offspring does not reach reproductive maturity. The hybrid may be sexually sterile like most mules (extremely rarely will a mule reproduce). The hybrid may be so different that it is unable to sexually interact with other representatives in its habitat. For example, some intraspecific hybrids of two species of frogs not only will look different from their parents, but more importantly the males will be unable to give an appropriate call and the females will not respond to calls from either parent species.

3. *Define each of the three speciation models.* 19.3, 19.4 There are three models of speciation. Allopatric (different/homeland) speciation occurs when there is an absence of gene flow between geographically separated populations. Physical barriers such as glaciers, mountain ranges, and bodies of water may keep them

apart. Daughter species may form gradually by divergence. If they remain separate long enough, enough differences may arise so that they are not able to interbreed. For example the formation of the Grand Canyon physically isolated two populations of rabbits. Now these populations have become so distinct that they can no longer interbreed and are considered different species.

In sympatric (with/homeland) species, the organisms involved have ranges that overlap. There is an absence of physical or ecological barriers, and the organisms may come in contact with each other. The textbook describes sympatric speciation among cichlid species of fish that inhabited two crater lakes in central Africa. Another example is the development of polyploidy in plants that immediately isolates the tetraploid plants from the diploid because triploid hybrids would be sterile.

Parapatric speciation occurs when two species share a common border. The two organisms may interbreed where they come in contact and produce highly variable species with some characteristics of both species. In Nebraska, a hybrid zone occurs where Baltimore orioles and Bullock orioles may mate. There has been a recent decline in interbreeding, and the two species are becoming distinct once again.

4. *Interpret these features of an evolutionary tree diagrams: a single line, softly angled branching, horizontal branching, vertical continuation of a branch, many branchings of the same line, a dashed line, a branch that ends before the present.* 19.5 An evolutionary or phylogenetic tree is a visual interpretation of how organisms arose and indicates their closest ancestors and relatives. Each branch represents one line of descent from a common ancestor. Each branch represents a line of speciation and divergence. Branches with small angles denote gradual changes occurring over long periods of time. Horizontal branches are examples of rapid evolution, as in the punctuated model. The vertical continuation of a line is an example of phyletic evolution. Gradual changes are incorporated through time until the new species is considered different from its progenitor. The occurrence of many branches from one point is an example of a precursor species giving rise to many new species. The presence of a dashed line indicates probable ancestry, and a branch that ends before the present time indicates extinction.

CHAPTER 20

THE MACROEVOLUTIONARY PUZZLE

1. *Will the fossil record ever be complete? Why or why not?* 20.1
The fossil record for the past has already been finished; that is, all the fossils of former organisms that will ever be preserved are now preserved. Some organisms that die now may eventually be fossilized. It is the nature of the fossil record that there are gaps because certain organisms did not die under conditions that would lead to their preservation. Those organisms with hard body parts tend to be preserved, while very few organisms with soft body parts are preserved. The actions of predators, scavengers, and decomposers destroy many forms before they become fossilized. Physical and chemical weathering will destroy many fossils. Fossils may not be found, or if found, they may not be recognized. Conversion of sedimentary rock into metamorphic rock will remove and/or destroy fossils. Many gaps exist in the fossil record for all of these reasons, plus the fact there were very few transitional forms ever alive. Because of this, we do not find a complete fossil record showing the origin of birds or man.

2. *Explain the difference between*
   a. microevolution and macroevolution C1
   b. homologous and analogous structures 20.4
   c. morphological divergence and convergence 20.4

(a) Microevolution refers to the small changes that occur in the gradual transition of an organism through time (such as the accumulation of differences that have occurred between a preserved fossil and its distant progeny that are alive today). Microevolution deals with evolution on a small scale and examines phenomena such as mutation, differential survival, genetic drift, selection, and speciation. The text defines microevolution as the changes in the allelic frequencies that have occurred over time. On the other hand, macroevolution studies the evolution of large taxa, such as the evolution of the birds or whales. Macroevolution refers to the patterns, trends, and rates of change among lineages over geologic time. (b) Homologous structures are similar because of a common ancestor. The forelimb of man and the forelimb of another vertebrate shows an evolutionary relationship. They may have bones with the same names, such as humerus, radius, and ulna. Analogous structures are similar because of a common function. The wing of a bird and the wing of a butterfly are similar in shape to allow flight but have no evolutionary connection.

(c) Morphological divergence results in departures in an ancestral form through selection. For example, the primitive pentadactyl limb has been modified to support different functions such as flying, swimming, burrowing, running, climbing, and so on. In convergence, two or more unrelated species develop similar structures. The similar fusiform body shape of the sharks, the ichthyosaurs, and the porpoises enabled these predators to swim rapidly through the water. The three have the same basic shape but represent three different classes of vertebrates.

3. *Give reasons why two organisms that are quite different in outward appearance may belong to the same lineage.* 20.4 Morphological differences in appearance do not necessarily distinguish different species. For example, in many species, sexual dimorphism may produce organisms that appear quite different even though they are the male and female of the species. Even Linnaeus erroneously classified the male and female of a duck as belonging to different species. Some species have some members that exhibit albinism (lack of pigment), rufescent (red coloration) or melanism (black pigmentation). These very striking members of the species differ from the rest of the species by virtue of the expression of a single rare allele.

4. *What mutations are the basis of a molecular clock and why?* 20.5 Neutral mutations are used to establish a molecular clock of evolution. A neutral mutation produces a change in the codons of the DNA specifying a protein such as cytochrome c̲. In some cases, the change may so alter the protein that it could not function in the hydrogen transfer system, and the organism possessing it would die. Many times the mutation would produce a substitution of a new amino acid for one in the primary structure of the proteins. If the alteration does not affect the functioning of the cytochrome protein, the mutation is said to be neutral. Through time, different substitutions accumulate. If they occur at a regular rate, then these

substitutions could serve as a molecular clock to establish the time frame for evolutionary change. If two animals' cytochromes were compared and there were three different amino acids in the protein, they could be said to have diverged millions of years ago. Another comparison indicating a difference of five amino acids would mean that the organisms being compared would have diverged earlier than those with just three differences. These comparisons add to our understanding of evolution and allow us to attach a time scale to evolution.

5. *Name a protein specified by a gene that has been highly conserved in organisms ranging from bacteria to humans.* 20.5 There are many conserved genes that are found in primitive forms of life such as yeast as well as in advanced forms including man. A typical example would be the gene controlling the protein cytochrome c in the electron transport system. Other examples would involve genes that control proteins that are basic to life processes, such as hemoglobin, chlorophyll, or vital enzyme systems.

6. *Why do evolutionary biologists apply heat energy to hybrid molecules of DNA from two species?* 20.5 It is possible to split DNA molecules of two organisms and place them so that the strands would base-pair with appropriate complementary base pairs on the other strands. They produce a DNA-DNA hybrid. The more complementary the two strands are, the more closely related are the DNA molecules. The base pairs of the DNA-DNA hybrid will separate when heat is applied. The amount of heat that has to be applied to separate the two DNA molecules is a measure of how much pairing has occurred. Higher levels of heat applied to achieve separation indicates the greater base-pairing and therefore similarity in the DNA molecules. Thus the amount of heat that has to be applied to a DNA-DNA hybrid provides a quantitative basis to compare the similarity of two DNA molecules.

7. *Define taxonomy. In what basic respect does cladistic taxonomy differ from classical taxonomy?* 20.6, 20.7 Taxonomy is the science of identifying naming, and classifying organisms. In cladistic taxonomy organisms are grouped according to similarities that are derived from common ancestry. In classic taxonomy, organisms are classified using a subjective analysis of differences between and similarities of different organisms.

8. *On the basis of recent evidence about a certain group of organisms, consensus has grown to change a five-kingdom classification scheme to a six kingdom scheme. Which group of organism is this, and what type of evidence favors putting it in its own kingdom?* 20.9 The organisms that have been added to the five kingdom classification scheme belong to the Archaebacteria. The archaebacteria were separated from the eubacteria in the Monera kingdom based upon the differences in their DNA sequences. The entire 1.7 million base pairs of DNA from an archaebacterium was sequenced and compared to members of the other kingdoms. The differences were vast and considered different enough to separate the two groups of bacteria into separate kingdoms.

UNIT IV EVOLUTION AND DIVERSITY

CHAPTER 21

THE ORIGIN AND EVOLUTION OF LIFE

1. *Compare the presumed chemical and physical conditions that are thought to have prevailed on   earth 4 billion years ago with conditions that exist today.* 21.1 Four billion years ago the earth was a molten mass of condensed gases and radioactive elements. The earth began cooling. Gases trapped below the earth's crust escaped to form the early atmosphere. The early atmosphere had little if any oxygen. Water did not exist in a liquid state because the heat at the earth's surface would have converted it back to atmospheric water vapor immediately. When sufficient cooling had occurred, precipitation started the hydrological cycle (rain, runoff and accumulation, evaporation), leading to further cooling of the earth. The earth's orbit was at a distance where water remained a liquid as opposed to a gas if it were closer to, or ice, if it were further away from the sun. Water in liquid form is a requirement for life to evolve. Through time the austere primitive conditions were ameliorated and environments capable of supporting live slowly developed.

Our knowledge of past conditions on earth are logical extrapolations of what earth must have been like according to the cosmic theory. Observations of Mars, the moon, and asteroids give added support for these conclusions. These conditions differ markedly from those that exist today.

2. *Describe examples of the kinds of experimental evidence for the spontaneous origin of (1) large organic molecules, (2) the self-assembly of proteins, and (3)the formation of organic membranes and spheres, under laboratory conditions similar to those of the early Earth.* **21.1, 21.2** Stanley Miller set up a system with two reservoirs, one containing liquid representing the ocean and the other containing gas representing the atmosphere. The two chambers were connected with glass tubing. The water in the "ocean" was boiled. In the atmosphere chamber four gases were present: methane, ammonia, water, and hydrogen. The gases were exposed to an electric spark to serve as an energy source to stimulate chemical reactions. After Miller allowed the system to operate for a week, he was able to collect from his system a number of organic compounds including amino acids. Similar experiments modifying the components of the atmosphere have demonstrated that nucleotides, lipids, sugars, and amino acids have been produced under abiotic conditions. ATP will even form when inorganic phosphate is added to the mixture. The intermediate stages of chemical evolution include the following: polypeptides, microspheres, lysosomes, and other semistable systems before the first cell evolved. These chemicals could be generated from their simpler precursors. Growth of crystals on different clays has been experimentally demonstrated. The latticelike features of clay could be used to manufacture long chains of polypeptides, some of which may have functioned as enzymes. Nucleotides could have been attracted to the clay lattice and functioned then as now as coenzymes. Sidney Fox heated amino acids under dry conditions to produce long polypeptides. These polypeptides or proteins were placed in hot water and allowed to cool. They formed microspheres. The microspheres will pick up lipids from the surrounding environment to form a lipid-protein film around each sphere. Liposomes have been experimentally produced.

3. *Summarize the key points of the theory of endosymbiotic origins for mitochondria and chloroplasts. Cite evidence that favor this theory.* **21.4** According to the endosymbiotic (inside/with/live) hypothesis proposed by Lynn Margulis, a large organism called a host species engulfed or incorporated a small prokaryote. Instead of digesting and destroying the guest species, the guest species was retained within a membrane in the body of the host species. The guest species had features that were very desirable to the host such as its ability to produce food through

noncyclic photosynthesis or to tap additional energy resources by its ability to conduct aerobic respiration and by its possession of additional cytochromes to use in the electron transport system. These became the eventual chloroplasts and mitochondria of eukaryotic cells.

Evidence to support the endosymbiotic theory includes the double membrane around the mitochondria and chloroplasts. The inner mitochondrial membrane resembles the plasma membrane of bacteria. The mitochondrion has its own DNA that divides independently of the host cell's DNA. The mitochondria DNA manufactures proteins (enzymes) used only in the mitochondria. The mitochondrial DNA code is slightly different from nuclear DNA. Chloroplasts also share the same features as the mitochondria. Chloroplasts show some structural variation that could be attributed to different lineages of photosynthetic eubacteria being incorporated inside eukaryotic forms.

4. *Describe the prevailing conditions that probably favored the Cambrian "explosion" of diversity among marine animals, as evidenced by the fossil record.* 21.5 The evolution that occurred during the Cambrian period took place in warm shallow seas that surrounded continents that were drifting apart. New marine environments developed that promoted adaptive radiation into new ecological opportunities. The explosive development of new forms were able to survive because there was a low-level of competition resulting from the availability of new adaptive zones. Novel experimental groups arose, some became extinct quickly while others flourished and diversified further. Conditions supporting biological experimentation continued until temperature lowered in the late Cambrian bringing about a massive global extinction.

5. *During which geologic time span did plants, fungi, and insects invade the land? What kind of vertebrates first invaded the land, and when?* 21.5 The first vertebrates to invade the land were the Labyrinthodont Amphibians that arose from the Crossopterygian fishes. These first amphibians arose in the Devonian (345–405 million years ago). The first vascular plants arose during the Silurian (405–525 million years ago). The insects made their entrance on the stage of evolution during the Carboniferous (345–280 million years ago).

6. *What were global conditions like when gymnosperms and dinosaurs originated?* 21.6 The reptiles gave rise to the dinosaurs whose dominance spanned the Mesozoic, approximately 175 million years. Their entrance onto the scene may have been prompted by a massive asteroid that impacted Quebec, Canada and their exit may be been the result of another asteroid impact near the Yucatan peninsular in Mexico. These cataclysmic events could have produced massive firestorms, volcanic eruptions, earthquakes and cold temperatures. Certainly, there would be massive extinctions clearing the way for the introduction of new forms of life - the reptiles at the beginning of the Mesozoic and the mammals at the beginning of the Cenozoic. The massive extinctions left many adaptive zones open for exploitation by the new organisms. At other times, during the Mesozoic there were major cataclysms that produced major extinctions and subsequent opportunities for adaptive radiation into the biotic voids resulting from the extinctions.

The flowering plants originated about 127 million years ago. 7 million years later the temperature rose by 25 degrees producing an environment that favored the rapid evolution and development of vegetation.

7. *Briefly explain how an asteroid impact and "global broiling" may have caused the mass extinction at the K-T boundary.* 21.7 The Alvarez hypothesis states that a massive impact of an asteroid near the Yucatan peninsula of Mexico led to the extinction of the dinosaurs. This event took place at the end of the Mesozoic period 65 million years ago at the boundary between the Cretaceous and Tertiary period (K-T). The global broiling theory proposed that the energy released at the K-T impact site was equivalent to the explosion of 100,00,000 nuclear bombs. Massive amounts of material was vaporized into a massive fireball. The events mimicked the impact of a comet that hit Jupiter in 1994. The immense heat must have destroyed all plant and animal life in large areas around the site. The persistent global cloud cover could have limited the available sunlight. Drastic climatic changes were induced including the development of an ice age. Severe climatic stresses may have brought about the extinction of many forms of life that survived the immediate effects of the impact.

8. *Would you expect the Paleozoic, Mesozoic, or Cenozoic to be called "the age of mammals"? As part of your answer, explain the differences between global conditions in each era.* 21.8 The Paleozoic would be called the age of the invertebrates, the Mesozoic would be the age of reptiles and the Cenozoic would be the age of mammals. Each of these geological eras were characterized by massive extinctions followed by adaptive radiation. The climates over the millions of years spanned by these eras would be expected to vary from mild conditions to times of glacier formation and colder conditions. One of the more important and often overlooked features that modified climate through time was the movement of continents that would shift the angle of insolation and the amount of energy available to control temperature. The process of orogeny (mountain formation) through earthquakes and vulcanism would alter wind patterns and create rain shadows and make major changes in the movement of air masses and pressure systems (storm cyclones) across the face of the land masses. Changes in ocean currents could have been a major factor in climate as we see in *El Nino* today.

CHAPTER 22

BACTERIA AND VIRUSES

1. *Label the structures on this generalized bacterial cell:* 22.1 The labels for the typical bacterium are as follows, starting from the five o'clock position and preceding clockwise: Cytoplasm, pilus, bacterial flagellum, DNA, ribosomes, cell wall, and capsules.

2. *Describe the key metabolic and structural features of bacteria. Make sketches of the three basic shapes of bacterial cells.* 22.1 Photoautotrophic bacteria carry on photosynthesis. Some bacteria are aerobic forms while others anaerobic. Chemoautotrophic bacteria usually get carbon from carbon dioxide and use organic compounds as a source of electrons and hydrogen. Others, called chemolithotrophs, use inorganic substances such as hydrogen, sulfur and nitrogen compounds as well as some forms of iron.

Photoheterotrophs derive their carbon from organic compounds such as fatty acids or complex carbohydrates from other organisms. The chemoheterotrophic bacteria derive their nutrients from living host (parasites) or organic products, wastes, or remains of dead organisms (saprobes). They form an extremely diverse group of organisms that fit a wide variety of niches and exhibit a multiplicity of metabolic life styles. They exhibit three basic shapes: coccus (a sphere), bacillus (a rod shape) and spirillum (comma-shaped or an elongated twisting spiral body). They may exist as separate individuals or clump together to form chains, sheets or grape-like clusters.

3. *Compared to your own bodily growth, how is bacterial growth measured.* 22.2 Human growth is measured by the accumulation of weight and increase in size (i.e. height). While bacteria may also increase in size, the more meaningful measure of growth is the increase in numbers resulting from reproduction of individuals to produce a colony. The rate of cell division in bacteria depends upon environmental conditions and food supply, but may increase to large populations in a very short time span.

4. *With respect to bacterial classification, what are some of the pitfalls of numerical taxonomy? How are comparisons of rRNA sequences from different bacterial groups assisting classification efforts?* 22.3, 22.6 Numerical taxonomy is the traditional method used to classify bacteria. The traits of an unidentified bacterium is compared to the characteristics of a known bacterial group. These traits include such traits as shape, motility, staining attributes of the cell wall, nutritional patterns, metabolic patterns, presence or absence of endospores, pigmentation structural features and others. The greater the similarity of the individual bacterial cell to the bacterial group to which it is compared, the more closely related the unidentified cell is to the bacterial group.

With the development of the capability to compare DNA through hybridization study, the analysis of gene sequencing and expansion of ribosome RNA, the process of classification has become more exacting. These analyses show that groups thought not be related by the numerical taxonomic approach are now demonstrated to be more closely related. The most significant development of the new genetic analysis is the demonstration that the archaebacteria are significantly different from the eubacteria. They are considered different enough that the recommendation is to elevate the archaebacteria to the status of a new sixth kingdom.

5. *Name a few photoautotrophic, chemosynthetic, and chemo-heterotrophic eubacteria. Describe some that are likely to give humans the most trouble medically speaking.* 22.4 Some of the photosynthetic bacteria include the cyanobacteria (formerly known as blue-green algae), prochlorobacteria (found in the tissues of invertebrates as a symbiont), and the purple or green bacteria in which oxygen is not a by-product of photosynthesis. The chemosynthetic bacteria include the nitrifying bacteria of the nitrogen cycle. Some representative heterotrophic bacteria include pathogens such as those that cause syphilis (spirochetes), Rocky Mountain spotted fever (rickettsia), pneumonia (mycoplasms), and strep throat (streptococcus). Other bacteria may obtain their food from various sources and may have economic importance. Pseudomonas cause swimmer's ear and could be a problem for those who enjoy water sports. Medically, any of the pathogenic forms such as those just listed would be potential disease producers.

6. *What is a virus? Why is a virus considered to be no more alive than a chromosome?* 22.7 A virus is a noncellular infectious agent possessing two characteristics: (1) it has a nucleic acid core surrounded by a protein coat and in some cases a lipid envelope; (2) a virus can replicate only after its genetic material enters specific host cells. Like a chromosome, a virus that is outside a cell exhibits no characteristics of life, but within a cell it is capable of replication and participates in transcription.

7. *Distinguish between*:
   a. Microorganism and pathogen C1
   b. Infection and disease 22.9
   c. Epidemic and pandemic 22.9

(a) A microorganism is a singe-cell organism that is too small to be seen without the aid of a microscope. They fill a variety of niches. Some microorganims are pathogens but not all pathogens are microorganims. Pathogens (disease/producer) are infectious disease producing agents that invade target organisms, multiply inside the host and may produce visual effects known as symptoms. The pathogens produces damage to the host's cells and disturbs its host.

An infection results when a pathogen invades the body of a host. A disease is the potential outcome of the invasion of the pathogen if the host's defense system is unable to overcome the pathogen's activities before they interfere with the host's normal functions.

An epidemic occurs when a pathogen spreads rapidly through a limited population for a limited time and then subsides health. Typical examples are the flu epidemic that become periodic health problems. A pandemic is a worldwide epidemic such as the AIDS pandemic.

CHAPTER 23

PROTISTANS

1. *Outline some of the general characteristics of protistans. Then explain what is meant by this statement. Protistans are often classified by what they are not.* C1 The protistans are the simplest and oldest and most primitive of eukaryotes. Protistans are most easily classified by what they are not because they are a highly variable group of organisms. The process of elimination is one of the ways to classify organisms and the protistans can be easily distinguished from the more primitive archaebacteria and eubacteria as well as from the members of the more advanced kingdoms: fungi, plants and animals.

The protistans reproduce by mitosis, meiosis or both. They exhibit 9+2 microtubule assemblies for both moving chromosomes and for cilia and flagella. Their DNA is associated with histone proteins. The photosynthetic forms use a wide variety of pigments and range from unicellular algae to giant seaweeds. The saprobes are similar to bacteria and fungi. The predators and parasitic forms are unicellular forms that resemble animals. Their cytoplasmic organelles are highly variable and specialized. The lines that demark the protistans from some of the other kingdoms are not clear. For example, some consider the large algae to be protistans while others classify them as plants. The slime molds are among the most exotic and enigmatic forms of life that have been discovered on earth.

2. *Review table 23.2, then cover it with a sheet of paper. Now name the major categories of protistans* . C1, Table 23.3 These organisms are eukaryotes, indicating that they possess a nucleus and a double-bound organelles such as chloroplasts and mitochondria along with other organelles. Most of them are unicellular so that their organelles and structural complexities produce a wide range of ultrastructure. The presence of chloroplasts enable them to produce food in the light. The euglenoids have light sensitive eyespots that allow them to move to the light. Locomotive organs such as cilia and flagella enable them to move and many exhibit positive and negative taxis so that they move to more favorable environments. The chytrids have enzymes suitable to digest organic material in their habitats. The phagocytotic slime molds are able to crawl over their substrates by amoeba-like movement to reach new food sources. Parasitic protozoans form cysts that enable them to survive adverse environments. The hardened shells of foraminiferans and radiolarians serve as protection. The trichocysts of paramecium may function in protection and food gathering. The contractile vacuoles help maintain osmotic balance enabling the organism to remove excess water. The presence of a gullet enables the paramecium to engulf its prey. Some protozoans have sensory cilia on their head or close to their mouths. Some of the multicellular algae are more complex and exhibit holdfasts (anchoring structures) stipes (stemlike parts), blades (leaflike parts) and bladders that serve as floats.

3. *Correlate some structural features of a protistan from each of the groups listed with conditions in their environments.*
   a . chytrids and water molds 23.1
   b. slime molds 23.1
   c. amoeboid protozoans 23.3
   d. animal-like protozoans and sporozoans 23.4, 23.5
   e. ciliated protozoans 23.6
   f. euglenoids, chrysophytes, and dinoflagellates 23.7
   g. red, brown, and green algae 23.8-23.10

(a) The chytrids are saprobic decomposers or parasites. They resemble fungal molds in their overall appearance and manner of nutrition. Some possess chitin as do the fungi. A newly germinated spore produces rhizoid-like absorbing filaments. Some multicellular forms have non-septate filaments (without cross walls) that are organized into a mycelium. Cytoplasmic streaming distributes nutrients throughout the body. The water molds are saprobic decomposers that derive nutrients from plant debris or necrotic tissue in living plants. Some are parasitic on aquatic animals such as those that infect aquarium fish. The water molds may develop extensive mycelia that may reproduce by gametes or spores. The downy mildew of grapes and the late potato blight that caused the Irish famines in the mid 1800s are typical representatives of the group. (b) The slime molds have amoeboid cells that crawl over rotten plant material feeding on bacteria, spores and organic compounds by phagocytosis. They can assemble into a motile slimy mass that releases reproductive spores and exhibits cytoplasmic streaming. (c) Amoeboid protozoans move by pseudopods. They engulf their food by phagocytosis. (d) The flagellates are groups of protozoa that move by means of flagella. They are free living predators or parasites that are particularly prevalent in the tropics. The sporozoans are non-motile parasites that can produce serious diseases. Perhaps the best known and most serious parasite is the one that causes malaria. (e) The ciliated protozoans are the most highly evolved and structurally complex of the protozoans. The well known *Paramecium* is a typical representative. It is covered by a flexible membrane called the pellicle that is covered by projections called cilia that provide motility to the protozoan. They have trichocysts that are miniature harpoons used for defense and to capture food. A gullet is present to capture food. A contractile vacuole periodically pumps out water to maintain water balance. There is a large macronucleus and a smaller micronucleus. (f) The euglenoid are flagellated organisms that have an eyespot capable of directing the protozoan to a lighted area where its chloroplast could carry on photosynthesis. Some are heterotrophic like other protozoans. The Chyrsophytes are represented by the diatoms. They have chlorophyll and a golden brown carotenoid, fucoxanthin. The dinoflagellates are armored marine phytoplankton. Some periodic blooms of these may color the surrounding water and produce a "red tide". These dinoflagellates release a neurotoxin that kill fish in the area and may kill many fish-eating birds. (g) The red algae are the delicate seaweeds that rarely exceed three feet in size. The common accessory pigment they possess is phycobilins which may mask their chlorophyll and alter their color. The brown algae are the large seaweeds often covering the rocks along a rocky seacoast. The giant kelps may be among the largest organisms in the world. They possess blades (leaf-like parts), stipes (stem-like parts), holdfasts (anchoring structures) and floats. They afford major habitats for marine life. The green algae are structurally and biochemically like the primitive plants. They are unicellular or filamentous or sheetlike. They produce starch grains and have cell walls composed of cellulose and pectin. They have chlorophyll a and b. Most are unicellular and inhabit freshwater environments.

4. *Select three different protistan species, then briefly explain how they adversely affect our affairs, such as crop yields and human health.* 23.1, 23.3, 23.4-23.7 Perhaps the most serious parasite the world has experienced is the malaria parasite, *Plasmodium*. It is a sporozoan that infects people all around the world, particularly in Africa. It is spread through the bite of an arthropod vector, the *Anopheles* mosquito. Humans and birds serve as hosts for the sporozoans. The parasite enters the bloodstream of the host and reproduce asexually in the liver and then invades red blood cells. Here they undergo cyclic asexual reproduction resulting in the typical chill and fever episodes characteristic of malaria. Another marauding female mosquito will pick up some of the sporozoans in the infected blood when she bites an infected person. The sporozoans will undergo sexual reproduction in the mosquito gut and then migrate in an infective form to the mosquito salivary gland. Perhaps 150,000,000 people become infected with malaria each year.

Between 1845 and 1860 the late potato blight destroyed the potato crop in Ireland. Over this period there were famines, deaths, and a massive migration brought about as a result of a water mold that devastated the potato crop. The climatic conditions lead to the spread of the disease. Perhaps there has never been a more destructive non-human pathogen to be ever visited on a country.

Dinoflagellates are one of the major phytoplankton found in marine environments. In the Tampa, Florida area massive runoff from rains wash phosphates from the phosphate mines on the west coast of Florida. At certain times of the year the additional nutrients mix with the appropriate climatic conditions to bring about a massive population explosion by the dinoflagellates. This bloom is called the "red tide" and the number of dinoflagellates is great enough to color the water. The dinoflagellates produce a nerve toxin that kill fish. The fish float on the surface and wash ashore. The dead fish represent a two prong attack on the important tourist industry. People don't want to share their beaches with dead fish and they are reluctant to go deep sea fishing with all the dead fish around. Other dinoflagellates may infect fish and shellfish that may poison humans that eat them.

5. *Select and briefly explain how the activities of photosynthetic protistan species have positive benefits for some communities of organisms, including human communities.* 23.7-23.10 The protosynthetic protozoans form the base for the major food chains in the ocean. The diatoms, desmids and dinoflagellates have sometimes been called the "grass of the sea". They and other photosynthetic forms are the producers in the marine ecosystem. Some of the large seaweeds form immense "forests" in the ocean that provide habitats and energy sources for the multiplicity of animals found around them. Some of the coral forming algae form and maintain some of the most exotic habitats for an immense variety of tropical fish. The freshwater phytoplankton contribute to the freshwater environment similar to the activities of the marine phytoplankton in the ocean. Humans primarily benefit indirectly from the photosynthetic protozoans although they directly utilize large amounts of agar, algin and carrageenan from marine organisms. The indirect benefits include the human position at the end of many of the food chains in the ocean, the diversity of life supported by these producers, the oxygen given up in photosynthesis as well as the reduction in atmospheric carbon dioxide also brought about by photosynthesis. These organisms participate in biogeochemical cycling and provide free services such as sewage disposal.

CHAPTER 24

FUNGI

1. *Describe the fungal mode of nutrition and explain how the structure of mycelia facilitates this mode.* 24.1 Fungi are usually colorless and lack chlorophyll so that they are unable to manufacture their own food. They are heterotrophs. If they derive their food from living organisms, they are parasitic, and if their source of food is dead, they are classified as saprophytes. In other cases, the fungus derives food from the body of an organism that was recently alive. These fungi are classified as decomposers, and they function to break down the body of dead organisms and recycle the elements found in them.

Often a fungal filament (hypha) will grow over or through the surfaces of its food source (substrate). The filament will release enzymes that will digest away the substance of the substrate, breaking the complex chemicals into simple soluble ones that are absorbed by the hyphae, transported through the fungus, and incorporated into its protoplasm. Other substances are simply broken down and carried away by flowing water or wind. These actions liberate the elements that were at one time locked in the substrate. These chemicals join the cycle of the elements so that they are again available for other processes as part of the biogeochemical cycle.

The mycelium is an amorphic mesh of individual fibers called hyphae. The mycelium is the body of the fungus. In some fungi, the mycelium becomes a definite body such as a mushroom, but in others (such as mildew). it appears to be spiderlike webs that invest whatever food source they encounter. The hyphae have cells walls reinforced with chitin. The mushroom is the reproductive body for the fungus.

2. *How does a lichen differ from a mycorrhiza?* 24.4 A mycorrhiza consists of a fungal mat of a basidiomycete that surrounds the roots of a higher plant. A lichen is an example of a symbiotic relationship in which the fungus derives food from an alga. The lichen is usually found above ground and is particularly

characteristic of tundra vegetation, while the mycorrhiza is a much larger, fungal relationship between the fungus and its plant host. It is estimated that eighty to ninety-five percent of all plants are infected with mycorrhiza. The mycorrhiza supplies the plant with water and minerals and in turn derives organic nutrients from the plants.

CHAPTER 25

PLANTS

1. *Identify a few of the structural and reproductive modifications that helped plants invade and diversify in habitats on land.* 25.1 About 400 million years ago, simple stalked plants invaded the terrestrial habitat, perhaps in symbiotic relationship with mycorrhizal fungi. Through time, many changes or trends developed, leading to the highly evolved terrestrial plants of today. Included in these changes were the development of vascular tissue, the development of dominance of the diploid sporophyte generation, the development of heterospory (spores of two types), the development of nonmobile gametes, the pollen grain providing the liberation from requirement of water for reproduction, plus the coevolution of flowers and insects and the development of seeds.

2. *Does the haploid phase or diploid phase dominate the life cycles of most plants.* 25.1 The diploid generation called the sporophyte generation dominates the life cycle of vascular plants. The haploid gametophyte generation becomes more insignificant in size and other characteristics as the process of plant evolution proceeded. In the most highly evolved group of plants, the angiosperms, the male gametophyte is reduced to three haploid cells in the germinated pollen grain and the female gametophyte to eight haploid cells in the embryo sac. The large visible dominant generation of vascular plants is the diploid sporophyte generation.

3. *Name representatives of the following groups of plants and compare their key characteristics: (also refer to table 25.1)*
   a. bryophytes and seedless vascular plants 25.2, 25.3
   b. gymnosperms and angiosperms 25.6, 25.8

(a) The Bryophytes include the liverworts, hornworts and mosses. They differ from the other plants in their lack of vascular tissue. In addition, the gametophytic generation was the larger dominant generation with the sporophytic generation smaller and mounted on and parasitic to the gametophyte. The seedless vascular plants include the "fern allies": whisk ferns (*Psilophyta*), club mosses (*Lycophyta*), horsetails (*Sphenophyta*) and ferns (*Pterophyta*). These plants developed vascular tissue-xylem and phloem. They were larger than the Bryophytes and had a dominant sporophyte generation, and were able to invade more habitats because they had vascular tissue. Some of the ferns and their allies became trees and were the dominant plants until the gymnosperms evolved. These giant tree ferns and their relatives were the source for our fossil fuels. The gymnosperms are generally evergreen conifers such as pines, hemlocks, fir, and spruce. Other less significant gymnosperms include the Cycads, Gnetophytes and the living fossil, Ginkgo. The flowering plants are the Angiosperms that are divided into two groups: Monocots and Dicots. The monocots include grasses, palms, irises, lilies, and orchids. The dicots include large numbers of herbaceous plants and woody shrubs and trees. Some of their well known representatives include the roses, sunflowers, legumes, mints, and deciduous trees. (b)Both gymnosperms and angiosperms are seed bearing plants, but the seeds of gymnosperms are borne unprotected on the scales of female cones while the seeds of angiosperms are protected by fruit. The chief advantage of the angiosperms is found in their flowers that attract pollinators and reduces the wasteful production of prodigious amounts of pollen that characterize the gymnosperms.

4. *Distinguish between:*
   a. *root system and shoot system* 25.1
   b. *xylem and phloem* 25.1
   c. *sporophyte and gametophyte* 25.1, 25.5
   d. *ovule and seed* 25.5
   e. *microspore and megaspore* 25.5

(a) The root system is usually found underground and provides anchorage, storage, and absorption of water and minerals. The shoot system is found above ground and functions to display the leaves to the sun for photosynthesis. (b) Xylem and phloem are the vascular tissue of a plant. Xylem is composed of tracheids and vessels. These cells are specialized tubes for the conduction of water and minerals. The two chief cells of the phloem are the sieve tubers and companion cells that function in the conduction of food. (c) The sporophyte is the dominant generation of vascular plants. Its cells are diploid and produce microspore mother cells and megaspore mother cells. The gametophyte generation is produced by the process of meiosis by which the microspore mother cells and megaspore mother cells generate haploid cells that develop into a gametophyte plant. In the vascular plants the gametophyte is reduced to a microscopic group of cells that produce the gametes (egg and generative nucleus found respectively in the embryo sac and pollen grain). (d) The ovule will develop into a seed after fertilization has occurred. The ovule is found within the ovary and contains the female gametophyte, the embryo sac. Seeds contain the embryo of the next sporophyte generation along with protective and nutritive tissues. Often the plants use the dormant conditions in the seed as a vehicle to avoid harsh environmental conditions and will only germinate when the environment becomes more favorable. (e) The microspore is the smaller of the two spores and is the male spore that develops into the male gametophyte or mature pollen grain. The megaspore is the haploid spore that develops into the female gametophyte embryo sac that produces the female gamete, the egg.

CHAPTER 26

ANIMALS: THE INVERTEBRATES

1. *List the six main features that characterize animals.* 26.1   The characteristics of animals are (1)They are multicellular and usually the cells are diploid and arranged in tissues, organs, and organ systems. (2) They are heterotrophs that derive their food from other living organisms. (3) They are aerobic and require oxygen for aerobic respiration. (4) They may reproduce sexually or asexually. (5) Most are motile, at least during part of their life cycle. (6) They exhibit stages of embryonic development.

2. *When attempting to discern evolutionary relationships among major groups of animals, which aspects of their body plant provide the most useful clues?* 26.1 One of the major features that distinguishes animals is the presence or absence of a vertebral column that divides them into two categories: the vertebrates and invertebrates. Other features include: (1) The type of symmetry (radial or bilateral), (2) the type of gut (one or two openings), (3) they type of coelom if present (no coelom, pseudocoelom, or true coelom), (4) presence or absence of segmentation and cephalization.

3. *What is a coelom? Why was it important in the evolution of certain animal lineages?* 26.1, 26.4-26.6 A coelom is a body cavity that occurs between the gut and the body wall of most bilaterally symmetrical animals. This plan is called a tube within a tube. Animals can be divided into three groups: (1) Acoelomates, without a body cavity, (2) Pseudocoelomates, body cavity lacks a continuous peritoneal lining, and (3) Coelomates, which possess a body cavity with a continuous peritoneal lining. Apparently, the availability of a peritoneal lining protected internal organs and allowed the development of larger, more complex animals.

4. *Name some animals with a saclike gut. Evolutionary, what advantages does a complete gut afford?* 26,1, 26.4, 26.6 The more primitive animals such as cnidaria and platyhelminthes had saclike guts, while the more advanced forms such as annelids, arthropods, molluscs, and chordates had digestive tubes or complete digestive systems. One simple advantage was that the

mouth was used for ingestion and the anus for egestion. Additionally, the complete digestive system is more complex and divided into regions where specific functions (such as digestion, storage, absorption of food) may occur. The availability of a complete gut enabled the development of larger size and complexity.

5. *Choose a species of insect that thrives in your neighborhood and describe some of the adaptations that underlie its success.* 26.15, 26.19 You could probably think of hundreds of successful forms, but the one I chose was the simple house fly. The development of complete metamorphosis was a major development—different lifestyles as a larva and an adult is an advantage. House flies obviously have a high reproductive rate. The ability to fly enables them to evade prey and search far and wide for food and mates. They are extremely quick and alert to threats from their environment. Their eyes with wraparound design and mosaic imaging make it hard to sneak up on them. They have an excellent sense of taste located in their legs. In many respects, the fly is one of nature's most interesting creatures. I would strongly recommend a delightful little book to all students: *To Know A Fly*, by Vincent G. Dethier, 1962, Holden Day Inc., San Francisco, CA, 119 pages.

CHAPTER 27

ANIMALS: THE VERTEBRATES

1. *List and describe the features that distinguish chordates from other animals.* 27.1 All chordates, both invertebrate and vertebrate, have a notochord (a supporting rod), a dorsal tubular nerve cord, a pharynx and gill slits in the pharynx wall, and a post anal tail. The notochord, gill slits, and tail may be found only in the embryo.

2. *Name and describe the features of an organism from both groups of invertebrate chordates discussed in this chapter.* 27.2 The two invertebrate chordates are the tunicates, or sea squirts, and

the lancelets. The tunicates are marine and are placed in the subphylum Urochordata because of the presence of a notochord in the tail of the swimming larva that resembles a tadpole. This larval form has the characteristic features of the chordates and to some degree also resembles the larvae of echinoderms. The larva provides us with the information to classify it because the adult bears no resemblance to it or other chordates. During metamorphosis, the notochord and tail disappear, and the animal attaches to the substrate and develops a tunic or covering made of a compound similar to cellulose. It becomes a filter feeder, with water being taken in through one siphon and being forced or squirted out another siphon (hence the name sea squirt). The cephalochordate, or lancelet, has a notochord extending from the head throughout the body. It has fins extending from the dorsal and ventral surfaces but is a poor swimmer because it lacks the paired fins of fish that allow coordinated swimming. It spends most of its life encased in a burrow with only its front end exposed. It is also a filter feeder.

3. *List four major trends that occurred during the evolution of at least some vertebrate lineages.* 27.3 The four trends that occurred during vertebrate evolution were (1) the development of a vertebral column to replace the notochord for support and locomotion; (2) the expansion of the nerve cord into a spinal cord and a brain. These will become the hallmark of the vertebrates; (3) development of gas exchange so that in the aquatic forms gills developed, while lungs replaced them in the terrestrial forms. Associated with these respiratory organs were modifications of the heart and circulatory system that transported the gases; (4) the development of coordinated locomotion. This involved paired fins in the aquatic forms, and with the development of skeletal supports in the terrestrial forms, the legs, wings, and arms formed.

4. *Which evolutionary modifications in fishes set the stage for the emergence of amphibians?* 27.3, 27.6 The preadaptations of the fish that allowed for the development of the Labyrinthodont Amphibians were the modification of the fleshy lobed fins into limbs that enabled their owner to crawl on land and the development and modification of lungs into an efficient system to obtain oxygen to allow for increased metabolism. In addition, there were changes in the sensory system to allow the amphibians a better awareness of their environment and the ability to catch

their prey. The lateralis system of fish (located along the middorsal line) enabled fish to detect vibrations. Chemical and vibrational (sound) receptors were transferred to the head region of amphibians.

5. *List some of the characteristics that distinguish reptiles from amphibians.* 27.7, 27.8 The reptiles differed from the amphibians by the development of the land egg. This was far superior and put the egg in an environment where it would less likely eaten. The reptiles developed scales that prevented desiccation and provided greater protection. The development of the penis as a copulatory organ by the reptiles meant that they did not have to return to water to reproduce. Reptilians kidneys conserved water. They also developed better locomotive abilities and became fearsome predators.

6. *List some of the characteristics that distinguish birds from reptiles.* 27.7, 27.8 Identifying characteristics of reptiles include the land egg with four extraembryonic membranes (amniotic egg); and the development of a copulatory organ, the penis; the reappearance of scales to prevent desiccation; and the development of increased complexity of all systems including the brain. Characteristics of birds include modification of scales to form feathers used for flight and insulation; development of a warm-blooded physiology that increased metabolic rates allowing flight; hollow bones; enlarged sternum with massive flight muscles; a four-chambered heart; an efficient respiratory system, including air sacs where gas exchange occurs while air flows through during both inhalation and exhalation. Birds still demonstrate their kinship to reptiles. They have scales on their legs, have a number of the same internal structures.

Characteristics of mammals include modification of scales to form hair or fur that is used for insulation; warm-bloodedness; mammary glands; development of differentiated teeth; and the development of the brain.

7. *List some of the characteristics that distinguish each of the three mammalian lineages from the reptiles.* 27.10, 27.11 The characteristics of mammals that distinguished them from reptiles include the formation of hair or fur that is used for insulation, warm bloodedness, mammary glands. They developed

differentiated teeth and different jaw structure. The hallmark of the mammals was the major development of the brain. Mammals often exhibit prolonged parental support for their young. Some mammals retain scales (i.e. mouse and rat tails) that proclaim their linkage to the reptiles.

The three mammalian lineages are the monotremes, marsupials and eutherians. The monotremes and marsupials retained archaic traits and resemble the reptiles more than the eutherians. The placental mammals were able to invade new adaptive zones by virtue of their greater ability to compete, their higher metabolic rate, more precise temperature regulation, and the use of the placenta and mammary glands to nourish their young.

CHAPTER 28

HUMAN EVOLUTION: A CASE STUDY

1. *What is the difference between "hominoid" and "hominid"? Are we hominoids, hominids, or both ?* 28.1 The word "hominoid" refers to apelike animals that resembled human ancestors as opposed to the more apelike forms. Hominoids were on a separate branch from the hominids. Hominids are restricted to those on the branch that directly led to the evolution of the humans. Humans are appropriately called hominids, hominoids, anthropoids, and primates.

2. *List the presumed macroevolutionary trends among certain primate lineages that were ancestral to modern humans.* 28.1 Some of the trends that were characteristic in the evolution of the primates are as follows: (1) change in overall structure and mode of locomotion; (2) modification of the hands, leading to increased dexterity and manipulation; (3) less reliance on sense of smell and more reliance on daytime vision, including color and depth perception; (4) change in dentition, toward fewer, smaller, less specialized teeth; (5) brain expansion and elaboration that led to the evolution of complex behavior. The major foundation for these trends was an arboreal life-style.

3. *What environmental changes are correlated with the great adaptive radiation of apelike forms during the Miocene?* 28.2 The movement of the land/masses through continental drift allowed for a greater dispersal but also produced cooler, drier climates that favored grasslands over forests. The change of the position of land masses and spread of grasslands allowed the radiation of apelike forms during the Miocene.

4. *On the basis of the fossil record, where did the first humans (of genus Homo) originate?* 28.3 In the late Miocene the first, *Homo habilis* evolved in the savannas of eastern and southern Africa.

5. *Why is it difficult to determine how australopiths were related and which ones might have been ancestral to early humans.* 28.2 The australopiths were evolved during a "bush" period of evolution when many ancestral lines originated, some died out while others may have integrated and shared characteristics. There are not an excessive number of fossils so that it is hard to piece together what was actually taking place. These organisms were placed together in a group called Australopithecus, the southern apes.

6. *Did H. Erectus populations coexist for a time with H. Habilis, which may have been ancestral to them.* 28.3 *Homo habilis* survived 500,000 years with very little change and remained in Africa. On the other hand, the populations of *Homo erectus* developed a more upright posture and migrated through Africa to Europe and Asia. They may have coexisted but *H. erectus* migration reduced any interaction between them.

7. *Briefly describe some of the conserved physical traits that link anatomically modern humans with their mammalian ancestors, then with their primate ancestors. What are the characteristics that set modern humans apart from other primates.* 28.3 Modern humans share some of the same physical characteristics with their more primitive ancestors such as specialized (although different) dentition, placental development of young, body hair and mammary glands. They share more specific traits with their primate ancestors such as binocular vision, increasing skull size, modification of the skull with the loss of the muzzle and modification of skeletal structure to produce bipedalism.

The primary differences between modern man and the primitive ancestors of humans could be summed up in the concept of cultural evolution. Modifications of behavior did not leave direct fossil evidence but the existence of tools and other indirect evidences provide clues of the social and cultural changes that were at the focus of human evolution. The minor physical changes that are recorded in the fossil record (i.e. small teeth and jaws) were far overshadowed by the development of cultural characteristics such as skill in communication. Consider the vast difference that separates an individual in small town USA with a member of an isolated tribe in South America to get some idea of the impact of cultural differences.

8. *Explain the difference between the multiregional model and African emergence model of modern human origins.* 28.4 There is no question that the process of human evolution started in Africa, but there are two alternative hypotheses to explain modern human origin. According to the multiregional model *H. erectus* spread throughout the world one million years ago. Isolated subpopulations responded to different environmental pressures to produce "races" in different places. These races differed phenotypically but still interbred as they do today. The African emergence model maintains that *H. sapiens* arose in Africa between 200,000 and 100,000 years ago. These population then moved out and either interbred or replaced the different racial population of the archaic *H. erectus*. Only then did regional phenotypic differences become superimposed on the original *H. sapiens* body plan.

UNIT V PLANT STRUCTURE AND FUNCTION

CHAPTER 29

PLANT TISSUES

1. *Choose a flowering plant and list some functions of its roots and shoots.* 29.1 The root system supports and anchors the plant, stores food, and absorbs and conducts water and minerals. The shoot system consists of the stem and leaves. The stem serves as the vascular connection between the roots and the leaves by transporting food, water, and minerals and displaying the leaves to the sun and flowers to pollinators. In some plants, the shoot system stores food. Additionally, both the shoot and root systems may synthesize a variety of organic compounds that have various uses.

2. *Name and define the basic functions of a flowering plant's three main tissue systems.* 29.1 The three tissues in vascular plants are 1) ground tissues including parenchyma, sclerenchyma, collenchyma cells used for food production, storage, protection, and support; (2) vascular tissues consisting of xylem and phloem used for transportation and, (3) dermal tissues composed of epidermal cells or their replacement, the periderm cells used for protection.

3. *Describe the differences between:*
    a. apical, transitional, lateral meristems 29.1
    b. parenchyma and sclerenchyma 29.2
    c. xylem and phloem 29.2
    d. epidermis and periderm 29.2, 29.8

(a) Apical meristems are located at the tips of stems and roots. The division of these cells leads to increases in length and produces primary growth. Some of the cells produced by the apical meristem form transitional meristems known as protoderm, ground meristem and procambium. These transitional meristems are a transitional or developmental stages that precede the formation of the epidermis, ground meristem and procambium. Lateral meristems are found inside stems and roots and are responsible for increases in width and for secondary growth. (b) Parenchyma is the most common type of cell found in ground tissue. These cells are relatively undifferentiated, with thin walls. They are metabolically active when mature. Sclerenchyma cells support mature plants and protect seeds. They have thick lignin-impregnated walls. Lignin strengthens and waterproofs cell walls, leading to the death of these cells. Some sclerenchyma cells are fibers, while others produce the grit characteristic of pear fruit. (c) Xylem and phloem are complex vascular tissues. Xylem cells include vessels and tracheids that are specialized for the conduction of water and minerals. These cells are dead and surrounded by lignin, which strengthens and supports the plant. Secondary xylem is known as wood. Phloem transports sugar and other solutes through its sieve tubes and companion cells. In secondary growth, the phloem, is found in the bark of trees.   (d) The epidermis is a single layer of dermal cells that surround and protect herbaceous plants. Often these cells are covered with a waxy waterproofing substance called cutin, which reduces the water loss from the plant. The peridermis replaces the epidermis in those plants with secondary growth. The majority of the peridermal cells are cork cells produced by the cork cambium. The cell walls of these cells are impregnated with a waterproof chemical called suberin. The cork cells are dead because they are unable to take in water or other substances. Historically, the first cells ever seen were cork cells.

4.   *Study figure 29.30. Is the plant that produced the yellow flower a dicot or monocot? Is the plant that produced the purple flower a dicot or monocot?* 29.2 The yellow flower is a dicot as indicated by a flower with floral parts in 5's and net venation in the leaves. The purple flower is a monocot with floral parts in 3's and parallel venation in the leaves.

5.   *Which of the following stem sections is typical of most dicots? Which of the following stem sections is typical of most monocots? Label the main tissue regions of both sections.* 29.2, 29.3 The stem on the left, with the vascular bundles arranged in a concentric circle, is a dicot, while the one on the right, with the smaller, more numerous veins, is the monocot. From the outside, the tissues are the

epidermis, parenchyma cells forming the cortex, vascular bundles (with phloem to the exterior and xylem to the interior), and a central pith in the dicot stem.

6. *Label the component of this three-year-old tree section. Correctly label the annual growth layers in the stem section below* 29.8 The labels on the side starting on top are as follows: bark, vascular cambium. The growth rings represent the first, second, and third growth rings.

CHAPTER 30

PLANT NUTRITION AND TRANSPORT

1. *Define soil in terms of its components. what is topsoil?* 30.1 Soil is the natural substrate upon which plants grow. It is derived from the physical and chemical weathering of the parent material formed from the rocks or crust of the earth. Mixed with these mineral-loaded fragments are variable amounts of decomposing organic matter or humus. Soil may be divided into three particle sizes called separates. The smallest are clay particles, the intermediate size are silt particles and the most coarse are sand particles. The texture of soil depends upon the relative mix of the three separates. Soils may be divided into different layers or regions known as horizons. The top horizon is called the A horizon and is known as top soil. It contains organic material that has been incorporated throughout the soil particles. Its depth is variable. It is the zone where plant roots extract nutrients from the soil solution found in the pores or spaces between the soil particles.

2. *Distinguish between:* 30.1
   a. Humus and loam
   b. Leaching and erosion
   c. Macronutrient and micronutrient (for plants)

(a) Humus is decomposing organic material found in the soil. Loam is a textural category of soil consisting of about equal proportions of

91

sand, silt, and clay. (b) Leaching refers to the removal of a soil nutrient that is dissolved in the soil solution as water flushes the nutrient through the soil. Erosion is the removal of surface layers of soil as a result of the movement of wind and water across soil. (c) Macronutrients are elements need in relatively high quantities (0.5% of the plant's dry weight) for normal plant growth. They include carbon, hydrogen, oxygen, nitrogen, potassium, calcium, magnesium, phosphorous and sulfur. Micronutrients include those elements needed in relative small amounts (i.e. parts per millions) and include chlorine, iron, boron, manganese, zinc, copper, and molybdenum. Although needed in smaller amounts, lack of these elements can generate poor health and produce nutrient deficiency symptoms.

3. *Refer to Table 30.1. What are some signs that a plant suffers a deficiency in one of the essential elements listed?* 30.1 Some of the symptoms of nutrient deficiency indicated in Table 30.1 are stunted growth, chlorosis (yellowing of leaves), pale green color, curled, mottled, or spotted leaves, burned leaf margins, weak roots and stems, deformed leaves, death of terminal buds, drooped leaves, discolored veins (purple), reduced reproductive growth, death of lateral branches, thickened or curled brittle leaves, leaf drop, necrotic (dead) spots, and other anomalies. In some cases, symptoms appear on older, lower leaves, other times on newly formed young leaves. Sometimes the veins are discolored, while other times the interveinal area is abnormal. The texture, shape, and general appearance may be different. Leaf margins may exhibit symptoms. The symptoms vary according to the particular nutrient that is missing.

4. *What is the function of Casparian strip in roots?* 30.2 The Casparian strip prevents water and dissolved substances from going around endodermal cells. It forces the dissolved substances to pass through the plasma membrane of the endodermal cells to reach the vascular cylinder of the root. Thus the plasma membrane and its transport proteins exert control over what reaches the vascular cylinder and therefore what is available to be distributed throughout the plant.

5. *Using Dixon's model, explain how water moves from the soil upward through tall plants.* 30.3 Henry Dixon proposed the cohesion-tension theory to explain how water reaches the top of a

plant. Water moves to the top of the plant to replace the water lost by transpiration. As water is removed by transpiration from the leaves, the water in the xylem is put under tension throughout the plant to the roots. As water is removed, the tension builds and pulls more water molecules up to replace the ones that are lost. The water in unbroken columns in the xylem coheres through hydrogen bonds that resist being pulled apart (the fact that the water is confined to the microscopic tubes of xylem allows the cohesion until the water molecules reach the leaves and transpire).

6. *Which type of ion influences stomatal action?* 30.4 The opening and closing of stomata is dependent upon the amount of water and carbon dioxide in the guard cells. A decrease in carbon dioxide levels in guard cells triggers active transport of potassium ions into guard cells. As the level of potassium in the guard cells increase an osmotic difference develops that favors the flow of water into the guard cells resulting in increase in the turgor of the guard cells and opening of the stoma. When the sun goes down the level of carbon dioxide rises and potassium and water leave the guard cell resulting in the stoma closing.

7. *Explain translocation according to the pressure flow theory described in this chapter.* 30.5 The pressure flow theory explains how solutes in the source region move to a sink region.
Photosynthesis in the leaves creates a source for sucrose. The sucrose moves into the sieve tubes of the leaves by active transport. The resulting high concentration of sucrose and low water potential causes water to flow into the area. The movement of water into the source region results in an increase in water pressure so that the water and dissolved sucrose move by bulk flow to a sink region such as the root or rapidly growing tissues. There the water will move out of the phloem into the sink cells. The sucrose is actively transported from the phloem to the sink cells, where it is used or converted into starch. Thus water enters the bulk flow system in response to the concentration gradient produced by active transport of sucrose in the source region.

CHAPTER 31

PLANT REPRODUCTION

1. *Define flower and define pollinator. What kinds of animals are attracted to flowers with red and orange components? What kinds respond to the flowers that smell like decaying organic material? What are some of the ways in which night-foraging animals find flowers in the dark?* C1 A flower is the reproductive organ that distinguishes the angiosperms from all other plants. A pollinator is any agent that functions in the transfer of pollen from the anther to the stigma. Pollinators include wind, water, insects and other animals such as bats and birds. Red and orange flowers attract birds and butterflies. Flies and beetles are attracted to flowers that have odors of decay. Night foraging animals are attracted to large white flowers with strong sweet odors. Animals are able to see the large white or pale flowers and use their sense of smell to detect their perfume.

2. *Label the floral parts. Explain the role each part plays in the reproduction of flowering plants.* 31.1 Starting at the top and proceeding downward the labels should be as follows: petal, anther + filament=stamen, stigma+stye+ovary=pistil or carpel receptacle. The petal serves as a colorful signal to attract pollinators. The anther is the reservoir where pollen grains are produced and stored. The filament is the stalk that supports and positions the anther in the flower. The stamen is the male sex organ of the flower. The stigma is the site for pollination. The style is the tissue between the stigma and the ovary. The compatibility of this tissue controls how well pollen tubes grow and determines if a pollen tube will reach the micropyle of the ovule. The ovary is the part of the flower that contains the ovules and will develop into the fruit after fertilization.

3. *Distinguish between these terms:*
   a. sporophyte and gametophyte 31.1
   b stamen and carpel 31.1
   c. ovule and ovary 31.1, 31.3
   d. microspore and megaspore 31.3
   e. pollination and fertilization C1, 31.3
   f. pollen grain and pollen tube 31.1, 31.1

(a) The sporophyte is the diploid generation that is dominant in the vascular plants. It reproduces by producing microspores and megaspores. The gametophyte is the haploid generation in the vascular plants. It is a microscopic generation that develops after the microspore and megaspore germinate. These gametophyte plants produce the male gamete (generative nucleus) and female gamete (egg). (b) The stamen is the male part of a flower and consists of the anthers and their supporting filaments. The carpel is the female part of a flower and consists of the stigma, style, and ovary. (c) Megaspore (large spore) is produced by meiosis and develops into the embryo sac, or female gametophyte. Microspores (small spores) are also produced by meiosis and develop into the male gametophytes or mature pollen grains. (d) Pollination refers to the transfer of pollen to the stigma. Fertilization involves the fusion of a sperm with an egg to produce a zygote. (e) The mature pollen grain is the male gametophyte and has three haploid nuclei, two generative nuclei, and a tube nucleus. The pollen tube forms when the pollen grain germinates. It forms a channel that the two generative nuclei use to reach the egg and polar nuclei within the embryo sac. (f) The ovule is the structure inside the ovary that is destined to become the seed after fertilization. It includes the nucleus, integuments, the stalks attached to the ovary walls, and the embryo sac. The female gametophyte consists of the eight haploid nuclei found inside the embryo sac. When the female gametophyte is mature it produces the female gamete, or egg. One of the eight haploid nuclei in the female gametophyte is the egg.

4. *Describe the steps by which an embryo sac, a type of female gametophyte, forms in a dicot such as a cherry (PRUNUS).* 31.3
Within the ovary small individual ovules start developing on the path toward the production of seeds. Within the ovule two protective layers grow around special tissue that undergoes meiosis to produce four haploid megaspores. Three of the megaspores

disintegrate. The remaining megaspore undergoes three mitotic divisions without cytokinesis to produce a cell with eight nuclei. The nuclei migrate to specific locations and then cell division occurs. There are seven cells formed. One cell retains two haploid nuclei that will fuse with one of the generative nuclei from the pollen grain to generate the triploid endosperm, the nutritive tissue for the future embryo. One of the other six cells functions as the female gamete, the egg. Note that since this tissue within the embryo sac is haploid and produces the female gamete it is the female gametophyte generation. The egg will fuse with the other generative nucleus from the pollen grain. Of course, the resulting zygote is diploid and the beginning of the diploid sporophyte generation. The zygote will develop into the embryo and the ovule will develop into a seed. When the seed germinates the potential exists to produce a new cherry tree.

5. *Define the difference between a seed and a fruit.* 31.4 A seed is a fully mature ovule that contains the embryo of the next generation. After fertilization the ovary of a plant will expand into a fully developed fruit containing one or more seeds.

6. *Do food reserves accumulate in endosperm, cotyledons, or both? If both, in what ways do these structures differ?* 31.4 In some plants, such as most monocots the reserve food material is restricted to the endosperm. In other plants, such as pecan the food material may be transferred to the cotyledons. The cotyledons are part of the embryo, the seed leafs and are diploid. The endosperm is triploid and functions only as a food reservoir.

7. *Name an example of a simple dry fruit, simple fleshy fruit, accessory fruit, aggregate fruit, and multiple fruit.* 31.4 Examples of simple dry fruit are: nuts, grains, achenes or legumes. Simple fleshy fruits include true berries (grape and tomato), drupes (peach, plum, cherry, olive) and hesperidium (citrus fruits). Raspberries are examples of aggregate fruits. Pineapples and figs are multiple fruits. See Table 31.1 for more details.

8. *Name the three regions of a fleshy fruit. What is the name for all three regions combined?* 31.4 The peach is an example of a fleshy fruit. The seed is surrounded by three layers of the pericarp. The external layer is the exocarp and is the protective skin of the peach. The middle layer is the mesocarp and is the fleshy part of

the fruit that is eaten. The innermost layer of the pericarp is the endocarp. It forms the pit that can be broken open to reveal the seed within. The peach is a simple fruit. Specifically, it is a drupe, a fleshy fruit with a pit containing the seed. Other examples would be cherries and olives. Apples are accessory fruits as well as being fleshy fruits. Specifically, apples are pomes. The distinction is that much of the fleshy tissue of the apple is made from receptacle tissue, while the fleshy tissue of the peach is composed of tissue from the ovary wall and is therefore part of the fruit (recall that a fruit develops from the ovary).

9. *Define and give an example of an asexual reproductive mode used by a flowering plant.* 31.7 Asexual reproduction in plants involves the use of plant parts to produce a new plant. This can occur naturally or may be used by man to reproduce certain species. Strawberries reproduce from runners. Grasses may spread from individual sprigs as new plants arise from nodes of horizontal stems. Underground storage organs can be separated and used to propagate plants. Examples include corms (gladiolas), tubers (potatoes) and bulbs (onions). Oranges, roses and other plants can be reproduced by parthenogenesis. Individual leaves can be used to propagate African violets and jade plants. Cuttings can be used to propagate shrubbery. Grafting is another method of vegetative propagation. Tissue culture is yet another method of asexual reproduction.

CHAPTER 32

PLANT GROWTH AND DEVELOPMENT

1. *Distinguish between plant growth and development.* 32.2 Plant growth simply refers to the increase in the number, size and volume of cells. Development refers to the emergence of specialized cells that have different structures and functions. Growth can be measured quantitatively (i.e. increase in weight, size, and length) while development involves qualitative features of cells.

2. *List the five known types of plant hormones and describe the known functions of each.* 32.2 Auxin (indole acetic acid): promotes elongation and growth, controls growth patterns. Gibberellin: overcomes genetic dwarfism, enables elongation of stems, breaks dormancy. Cytokinin: promotes cell division, retards aging. Abscisic acid: participates in opening and closing of stomata, triggers dormancy. Ethylene: promotes fruit ripening.

3. *Define plant tropism and give a specific examples.* 32.2 A plant tropism is a growth response of a plant to a unidirectional stimulus. The tropism is termed positive if the plant organ grows toward the stimulus and negative if it grows away from the stimulus. The prefix used with tropism indicates the environmental trigger such as photo=light, gravit=gravity, hydro=water, thigmo=touch, rheo=current, chemo=chemical.

A simple example involves thigmotrophic response of a grape vine to an arbor support. The stem comes in contact with the arbor support. Auxin accumulates on the side of a stem or tendril opposite the point of contact with the arbor support. The auxin stimulates the outer side of the stem to grow faster producing a curve to the growth of the stem. More of the stem then comes in contact with the arbor support stimulating more auxin to accumulate on the outer surface which causes the exterior of the stem to grow faster eventually causing the stem(or tendril) to wrap around the support.

4. *What is a phytochrome and what role does it play in flowering process?* 32.4 Phytochrome means plant color. Phytochrome is a blue-green plant pigment that responds to red light and is turned off by far-red light waves. It functions as a biological clock. The phytochrome pigment is converted into its active form (Pfr) by the red colors that dominate the sky at dusk. The phytochrome is converted to its inactive form (Pr) in darkness. Pfr controls seed germination, stem elongation, leaf expansion, and flowering. It controls the photoperiodic response of flowering.

Photoperiodism is the biological response of organisms to changes in the relative length of light and dark periods. It is one of the major controls over the timing of biological processes in nature (i.e. flowering, dormancy, migration, hibernation, estrus cycles, etc.). The most widely studied biological photoperiod response is control over flowering. Plants are classified as long-day, short-day or day-neutral plants.

Long-day plants bloom in the late spring and summer when the day length exceeds a critical time period. Short day plants bloom in the fall when the dark period exceeds a critical length (if the dark period is interrupted by a short light period the plants will not be induced to flower). The biological clock controlling this response is the phytochrome molecule. Phytochrome has an inactive form called Pr and an active form called Pfr. When sufficient active form of phytochrome accumulates the molecule Pfr triggers the appropriate photoperiodic response.

UNIT VI ANIMAL STRUCTURE AND FUNCTION

CHAPTER 33

TISSUES, ORGANS, AND HOMEOSTASIS

1. *Describe the characteristics of epithelial tissue in general. Then describe the various types of epithelial tissues, in terms of specific characteristics and functions.* 33.1 Epithelial cells come from layers of closely packed cells that adhere to one another to form a barrier that prevents invasion of the body by foreign organisms. They not only cover the entire external surface of the body, but they also line the openings to the body and the lumen of the digestive tract. They may be a single layer of cells, described as simple epithelium, or they may have more than one layer and be called stratified. They may have their free surface covered with cilia to clean and filter the air in the respiratory tract or to create currents in the oviducts to carry the ovum toward the uterus. Some absorptive cells in the intestines have finger-like projections called microvilli that increase the absorptive surface. Underlying the epithelium, before the connective tissue, there is a basement membrane rich in proteins and polysaccharides.

The glandular epithelium are cells of both exocrine and endocrine glands that produce and secrete enzymes, hormones and other products such as mucus, oil, sweat, saliva, earwax, and milk. Cuboidal epithelial cells lining the tubules of the kidney function in filtration, reabsorption, and secretion. Germinal epithelial cells will eventually undergo meiosis and participate in gametogenesis. The epithelial tissues of the body function in protection, absorption, secretion, excretion, and reproduction.

2. *List the major types of connective tissues; add the names and characteristics of their specific types.* 33.2 Connective tissue may be either loose or dense connective tissue. The matrix of loose connective tissue contains collagen and elastic fibers and fibroblast cells that produce the fibers and the ground substance. Dense, irregular connective tissue, as its name implies, has more and thicker fibers that serve to protect organs such as the testis. There are specialized connective tissues, such as dense, regular connective tissue that forms tendons and ligaments. Cartilage and bone are skeletal elements in the vertebrates. The characteristics of their matrix gives them flexibility and resilience so that they are excellent supporting and protecting tissues. Adipose tissue is a specialized connective tissue that stores fat. The last type of connective tissue is blood, which has a liquid matrix and carries three types of cells: red blood cells, while blood cells, and blood platelets.

3. *Identify and describe the following tissues:* 33.1-33.3 The photographs are as follows: top left-simple cuboidal tissue; top right-loose connective tissue; bottom left-skeletal muscle; bottom right -adipose tissue.

4. *Identify this category of tissue and its characteristics.* 33.4 The photograph illustrates nervous tissue.

5. *What type of cell serves as the basic unit of communication in nervous systems?* 33.4 The cell that serves as the basic unit of communication in the nervous system is the neuron.

6. *Define animal tissue, organ, and organ system. List and define the functions of the eleven major organ systems of the human body.* C1, 33.6 A tissue is a group of cells and intercellular material with a common structure and function. An organ is a structural unit in which tissues are combined in definite proportions and patterns that allow them to perform a common activity. Organ systems consist of two or more organs that interact chemically, physically, or both in such ways that they contribute to the survival of the organism. The major organ systems of the human body and their functions are as follows: (1) integumentary system covers and protects the body from injury and dehydration, exerts temperature control, excretes some wastes, receives some external stimuli, and provides defense against microbes; (2) muscular system is

responsible for movement of the whole body and its internal parts, the maintenance of posture, and the production of heat; (3) skeletal system provides protection of body parts, sites for muscle attachment, production of blood cells, and storage of calcium and phosphorus; (4) nervous system detects external and internal stimuli, controls and coordinates responses to stimuli, and integrates the response of all organ systems; (5) endocrine system controls body function through hormones in concert with the nervous system; (6) circulatory system allows for rapid internal transport of gases, foods, wastes, hormones, and other substances to and from cells and helps regulate internal pH and temperature; (7) lymphatic system circulates and returns some tissue fluid to blood and serves in the body's immunity system; (8) respiratory system provides cells with oxygen and removes carbon dioxide as well as participating in pH control; (9) digestive system ingests, digests, and absorbs food and egests undigested wastes; (10) urinary system maintains the volume and composition of extracellular fluid and the excretion of blood-borne wastes; and (11) reproduction system in males involves the production and transfer of sperm, while the female produces eggs and an environment that protects and nurtures the developing embryo and fetus. All systems interact to provide for homeostasis, which enhances the organism's chance for survival.

7. *Define extracellular fluid and interstitial fluid, and plasma.* 33.7 Extracellular fluid refers to the fluid located outside of cells. Much of this fluid is called interstitial which means that it is found in the spaces between cells and tissues. Plasma is the fluid component of blood.

8. *Define homeostasis.* C1, 33.7 Homeostasis means maintaining the status quo. In physiology, homeostasis refers to the processes involved in maintaining a stable internal environment.

9. *Briefly describe two major categories of the homeostatic mechanisms operating in the human body .* 33.7 In order to maintain homeostasis within the body there must be receptors that constantly monitor internal conditions, integrators that evaluate conditions, and effectors that can change conditions. There is a set point, such as the glucose or pH level or oxygen concentration in the blood, the body temperature, or some other internal factor. As the internal conditions change as a result of metabolism, a biological feedback system is turned on to bring the conditions back to the set

point. The glucose level could be corrected by the secretion of either insulin or glucagon. Once the glucose level reverts back to the set point, the secretion of the hormone will cease until conditions change enough for it to be secreted again. This system is similar to a thermostat that controls a heater and air conditioner and is known as a negative feedback system. The second control system is a positive feedback system in which a physiological response is intensified as occurs in the uterus where oxytocin secretion results in labor and eventually childbirth.

CHAPTER 34

INFORMATION FLOW AND THE NEURON

1. *Define sensory neuron, interneuron, and motor neuron in terms of their structure and functions.* 34.1, 34.4 A sensory neuron carries impulses from a sense organ to the central nervous system. An interneuron is located in the brain or spinal cord. It receives an impulse from a sensory neuron, integrates the information, and then influences other interneurons or motor neurons. A motor neuron carries an impulse away from the central nervous system or interneuron to an effector, such as a muscle or a gland.

2. *Label the functional zones of this motor neuron.* 34.1 The input zones are to the left of the diagram and include the dendrites and the cell body. The output zones are to the right and include the ends of the axon.

3. *Define resting membrane potential, graded potential and action potential.* 34.1 Then a nerve exhibits resting potential and it is charged and ready to respond to a stimulus. A steady voltage difference is maintained across the plasma membrane of a nerve cell that is not transmitting a nerve impulse. An action potential is a quick depolarization and repolarization of a neuron in response to a stimulus and results in the transmission and propagation of a nerve impulse. Graded potentials are local stimuli that are at sub-threshold levels. Two or more of these may be added together to reach the threshold level and produce an action potential. The

local signals (graded potential) may reach a "trigger zone," an area of the membrane perforated with many ion channels. If the local signal reaches this area it may be able to induce an action potential. Action potentials are usually induced by stimuli that exceed the threshold level. The availability of trigger zones allows stimuli that are at subthreshold levels (known as graded potentials) to be added together to produce an action potential.

4. *A neuron at rest is controlling the ion distribution across its plasma membrane. Identify the two major kinds of ions. Do they leak across the membrane, are they pumped across, or both? As part of your answer, describe gated and ungated transport proteins embedded in the neural membrane.* 34.1 The two major ions involved are sodium and potassium. Some of the protein channels are always open while others only open during the propagation of an action potential. There is some slow leakage through the open protein channels so that energy must be expended to operate the sodium-potassium pump to maintain the concentration differences established across the neural membrane during the resting potential. If the sodium-potassium pump did not operate, the voltage difference would eventually dissipate and disappear and it would not be possible to generate an action potential. Transport proteins are channels through the neural membrane for the passage of ions. The ungated transport proteins are always open and allow a continual leakage of ions across the membrane. The gated transport proteins open and shut in response to the passage of action potentials. Stimulation may produce a cascade of sodium ions through a gated channel to the interior of the neuron producing a wave of depolarization. Another gated channel may influence the flow of potassium ions to end an action potential and reestablish the resting potential.

5. *With respect to action potentials, explain what is meant by threshold level, by all-or-nothing spikes, and by self-propagation of an action potential.* 34.2 At the resting potential, a voltage differential of about -70 millivolts is maintained across the neural membrane. A single stimulus has to exceed -70 millivolts to initiate an action potential. Graded signals can spread from an input zone to a trigger zone. If the sum of graded signals exceed -70 millivolts the threshold has been reached and an all-or-none spike is initiated through a positive feedback cycle. The all-or-nothing

message refers to two things: (1) the stimulus must exceed the threshold level before an action potential can be initiated, and (2) once an action potential is initiated it will be propagated throughout the total length of the neuron. Sodium ions flow through gated channels for about a millisecond. The flow of sodium decreases the negative charge inside the neuron causing other gates to open and more sodium to enter so that the charge along the membrane reverses. When the charge becomes positive the potassium ions flow out of the neuron and voltage is restored. The changes in electrical charges that occur in an input region will trigger changes in adjacent areas, thereby propagating the action potential along the nerve in a process of self-propagation. After the action potential travels down a neuron, the sodium potassium pump reestablishes the ion gradients so that the neuron is ready to produce another action potential.

6. *Define chemical synapse and neurotransmitter. Choose an example of a neurotransmitter and state where it acts.* 34.3 A chemical synapse is a narrow junction between the output zone of an initiating (presynaptic) neuron to the input zone of an adjacent receiving (post synaptic) neuron. A neurotransmitter enables the action potential to bridge the gap between the two neurons and allow the action potential to continue down the second neuron. Neurotransmitters diffuse across the small space of the synapse between the first nerve and a second nerve or a muscle or gland cell. Acetylcholine (ACh) is a neurotransmitter that acts at the chemical synapse between a motor neuron and a muscle cell. The ACh released from the presynaptic neuron diffuses across the space to bind on receptors of the muscle cell membrane. The ACh triggers an action potential and causes the muscle to contract.

7. *Define synaptic integration. Include definitions of EPSPs and IPSPs as part of your answer.* 34.3 Synaptic signals are graded potential (local signals) at the synapse. There are two types: (EPSP) excitatory postsynaptic potentials in the postsynaptic potentials that depolarize or stimulate the postsynaptic cell, and (IPSP) inhibitory postsynaptic potentials that hyperpolarize or depress the post synaptic cell. In synaptic integration the competing or complementary signals are summed to determine whether the threshold is reached and if an action potential will be generated. This interaction between EPSP and IPSP determines whether a signal arriving at a neuron will be dampened,

suppressed, reinforced or sent onward to other cells in the body. Such ability is important to controlling the coordination of body responses to stimuli.

8. *What is a myelin sheath? Do all neurons have one?* 34.4   A myelin sheath is a lipid covering that surrounds the axons of some sensory and motor nerves. It is secreted by Schwann cells and speeds up the rate of transmission of those neurons that are encased by myelin. The transmission of an impulse may jump from spaces between the Schwann cells known as nodes thus speeding up transmission. The ascending and descending tracts in the spinal cord are surrounded by myelin sheaths to allow rapid transmission to and from the central nervous system.

9. *How do divergent, convergent, and reverberating circuits among blocks of neurons differ from one another?* 34.4   Regional blocks of neurons receive excitatory and inhibitory signals that they that they integrate and send out new signals in response. Divergent blocks send their signals out in many directions. Convergent blocks receive signals from many sources and send them on to a few. Reverberating blocks receive signals and send them back on themselves to repeat the signals over and over again.

10. *Define reflex, then give an example of a reflex arc.* 34.4   A reflex is the simplest of all nerve responses. It involves a sensory neuron, an interneuron and a motor neuron. The stimulus does not have to reach the brain before it is acted upon. For example, if you picked up a hot iron the stimulus would travel from a sensory neuron to the spinal cord where an interneuron would transmit it to a motor neuron that would stimulate muscles to release the iron and pull away. Some time after the action the brain would become aware of the pain.

CHAPTER 35

INTEGRATION AND CONTROL: NERVOUS SYSTEMS

1. *Generally describe the type of nervous systems found in radially symmetrical animals and in bilaterally symmetrical animals.* 35.1 In radially symmetrical forms, all parts of the environment share equal importance. There is no front or back. The nervous system is a simple nerve net that supplies all parts of the body equally and is responsive to stimuli in all directions. It is said to have a decentralized nervous system. On the other hand, bilaterally symmetrical forms are oriented front to back and exhibit cephalization, a concentration of sense organs and ganglia or brain in the head region. There are paired nerves to each side that extend to the back end or tail of the animal. They provide information to the central nervous system and allow for responses in a coordinated fashion. Bilaterally symmetrical forms became highly successful predators and are capable of escaping from other predators.

2. *What constitutes the central nervous system? The peripheral nervous system?* 35.2 The central nervous system consists of the brain and spinal cord. The peripheral nervous system consists of the cranial and spinal nerves and ganglia.

3. *Distinguish between the following:*
    a. *central and peripheral nervous system* 35.2
    b. *cranial and spinal nerves* 35.3
    c. *somatic and autonomic nerves* 35.3
    d. *parasympathetic and sympathetic nerves* 35.3

(a) The central nervous system consists of the brain and the spinal cord, while the peripheral nervous system consists of cranial and spinal nerves and their ganglia. (b) The cranial nerves are the twelve pairs of nerves that originate in the brain, while the spinal nerves are 31 pairs of nerves that originate from the spinal cord. (c) The somatic nervous system consists of nerves leading from the central nervous system to muscles, while the autonomic system is comprised of nerves leading to the visceral part of the body (heart, digestive organs). The somatic system is under voluntary control, while the autonomic system is not under voluntary control. (d) The parasympathetic nervous system originates from the cranial and sacral regions of the spinal cord, tends to slow down pulse rate, and promotes normal "housekeeping" functions, such as digestion. The sympathetic nervous system arises from the thoracic and lumbar regions and stimulates that part of the body to prepare it for fight or flight.

4. *Distinguish between:*
    a. *White matter and gray matter* 35.2
    b. *Nerve and tract* 35.2

(a) The white matter is that portion of the spinal cord which contains the ascending and descending tracts. It is white because the nerves are covered with a myelin sheath to speed conduction of impulses. The gray matter is composed of the dendrites and cell bodies of neurons and neuroglial cells in the cerebral cortex and the butterfly-shaped central portion of the spinal cord. (b) Nerves are cord-like connections of the peripheral nervous system that are composed of the axons of sensory neurons or motor neurons or both. In the brain and spinal cord similar cord-like bundles are called tracts.

5. *Describe the structure and function of the spinal cord.* 35.3 The spinal cored is the connection between the peripheral nervous system and the brain. It allows for rapid communication of sensory information to the central nervous system and instructions from the central nervous system to effectors throughout the body. A reflex arc is the site where direct connections are made between sensory and motor neurons via interneurons. The spinal cord runs through the central canal in the vertebrae in the backbone. It is covered by a tough three-layered membrane called the meninges that protects and isolates the spinal cord. There is a central butterfly-shaped gray area surrounded by a the white area which contains myelinated nerve fibers for rapid transmission of impulses via the ascending and descending tracts.

The spinal cord is essentially an extension of the brain that extends from the base of the brain to the second lumbar vertebra.

6. *Label the major parts of the human brain.* 35.4, 35.5 The labels, starting from the top right (one o'clock position) and proceeding clockwise are as follows: cerebral cortex. location of pineal gland, cerebellum medulla oblongata, pons, one of the two optic nerves, hypothalamus, thalamus, and corpus callosum

7. *Explain the difference between the reticular formation and the limbic system in terms of their component parts and main functions.* 35.4, 35.5 The reticular formation is found in the thalamus. It consists of a major network of interneurons in the brain that monitors and controls the activity of the nervous system. If you were awakened in the middle of the night by a strange noise your reticular formation would alert your sensory system to identify the noise. The limbic system is the part the brain that controls the emotions. The limbic system is connected to the olfactory lobes and the cerebral cortex as well. The hypothalamus serves as the gatekeeper of the limbic system.

8. *Define cerebrospinal fluid and the blood-brain barrier.* 35.4 The cerebral spinal fluid is a clear fluid secreted by the four ventricles (cavities in the brain). It serves to protect and cushion the brain. The blood brain barrier is a set of mechanisms that help to control which blood-borne substances reach the neurons of the brain.

9. *Briefly describe one of the brain centers of the cerebral cortex; mention the tissue lobe in which it is located.* 35.5 The visual center is located in the occipital lobe. The occipital lobe is located at the lower back of the head.

10. *Distinguish between:* 35.7
   a. *Short-term memory and long-term memory*
   b. *Fact memory and skill memory*

(a) The words indicate the type of memory involved. Short-term memory involves memory of events over the span of a few hours. For example, what you had for breakfast or whether or not you had brushed your teeth. Long-term memory involves a much longer time span such as what you got for your sixth birthday and who came to the birthday party. (b) Explicit information such as dates, telephone numbers, names, odors, musical terms are facts that may be forgotten or committed to long term memory. Skill memory deals with the ability to perform certain motor activity. Your skill increases with practice. Such activities as riding a bicycle, skating, dancing, gymnastics involve skill memory.

11. *Which part of the brain has major influence over states of consciousness, which include sleeping and arousal.* 35.8 The control over consciousness is located in the reticular formation. This portion of the brain can cause you to become super alert to your surroundings or produce lack of awareness, drowsiness and sleep. Substances its secretes includes serotonin that inhibits wakefulness.

12. *What is a habit-forming drug? Can you describe the effects of one such drug on the central nervous system?* 35.9 Habit forming drugs include stimulants, depressants, hypnotics, analgesics, psychedelics, and hallucinogens. The most widely abused of these substances is ethyl alcohol. It acts directly on plasma membranes to alter cell function. As the amount imbibed increases, there is a depression of brain functions from front to back. Symptoms include disorientation, slurring of speech, dizziness, loss of balance, and double vision. If more alcohol is taken, further brain centers are blocked until unconsciousness and even death may occur. Long-term usage can lead to cirrhosis of the liver. Habit forming drugs exhibit both psychological and physiological addiction.

CHAPTER 36

SENSORY RECEPTION

1. *What is a stimulus? When sensory receptors detect a specific stimulus, what happens to the stimulus energy?* 36.1 A stimulus is any type of energy that the body can detect through its sense receptors. Although we usually think in terms of external stimuli, such as sound or odor, many stimuli are internal. Often the individual cells that perceive the environment are organized into sense organs that amplify or focus the energy into an action potential that travels along specific pathways to designated areas of the brain.

2. *Name six categories of sensory receptors and the type of stimulus energy that each kind detects.* 36.1

CATEGORY	STIMULUS
(1) Mechanoreceptors	Touch and pressure, stretch vibration, (sound waves), fluid movement
(2) Thermoreceptors	Temperature change (heat, cold)
(3) Nociceptors (pain receptors)	Tissue damage, burns
(4) Chemoreceptors	Taste, smell, chemicals
(5) Osmoreceptors	Change in solute concentration
(6) Photoreceptors	Light

3. *How do somatic sensations differ from special senses?* 36.1
Somatic sensations are detected by sensory receptors scattered throughout the body as opposed to special senses that are activated neurons found in specific sense organs such as the eye, ear, and nose. The somatic sensations are fed to the primary somatosensory cortex, a strip above the ear that spans the upper surface of the cerebrum from one side to the other. Sensations such as pain, touch, pressure, cold, and warmth are detected by isolated receptors. The kinesthetic sense is based upon stimulation of proprioreceptors scattered through the body. This gives the ability to know where body parts are even in the dark so that you can tie your shoe.

4. *What is pain? Describe one type of pain receptor.* 36.2 Pain is a sensation of discomfort associated with injury to a body part or damage to cells. Nociceptors are free nerve endings that detect stimuli that cause tissue damage. The signals from the nociceptors are relayed to the parietal lobe of the cerebrum, where they are processed.

5. *Describe the properties of sound. Which evolved first, the sense of balance or sense of hearing? Do both senses require participation of the outer, middle, and inner ear?* 36.4, 36.5 Sound is a form of energy that moves in a wavelike pattern similar to ripples formed when a rock is dropped into water. The waves behave as vibrations passing through the air to the ear. Sound waves have different intensities that depend upon the amplitudes of the sound waves-large amplitudes produce greater sound. Sound waves vary in frequency thereby controlling the pitch. Long waves

with low frequencies produce low tones while waves with high frequencies produce high pitched sounds. The inner ears of animals such as fish, amphibians and reptiles first evolved as organs for equilibrium. The sense of balance is designed to monitor the position of an animal with respect to gravity and can detect motion and acceleration to give the animal an awareness of its position in the environment. The sense of balance is located in the part of the inner ear known as the vestibular apparatus. It consists of three semicircular canals and the cavities: utricle and saccule. The sense of hearing involves all three parts of the ear. The outer ear collects the sound waves. The middle ear consists of three small bones that transmit vibrations to the inner ear. The cochlea of the inner ear translates the vibrations that reach it into nerve impulses that are interpreted in the temporal lobe of the cerebrum as sound.

6. *How does vision differ from light sensitivity? What sensory organs and structures does vision require?* 36.4, 36.5 Almost all organisms are sensitive to light and dark. Vision requires eyes and a complex neural pattern in the brain that can interpret the patterns of light and dark through the action potentials generated by the photoreceptors. The incoming signals encode information about position, shape, brightness, distance, and movement of visual stimuli. In some cases, the photoreceptor also senses color. The photoreceptors that are responsible for these functions are found in the eye.

7. *Label the component parts of the human eye.* 36.7 The parts of the eye, labeled in a clockwise fashion beginning at the upper right, (at the one o'clock position) are as follows: retina, fovea, optic disc (blind spot), optic nerve, vitreous humor, ciliary muscle, aqueous humor, cornea, pupils, lens, iris, choroid, and sclera.

8. *How does the vertebrate eye focus light rays from a visual stimulus? What do nearsighted and farsighted mean?* 36.7, 36.9 Ligaments attached to the ciliary muscle change the shape of the lens and focus the light rays from the image onto the retina. When the ciliary muscle contracts, the lens bulges, and the focal point moves closer to the lens. When it relaxes, the lens flattens, and the focal point moves further back toward the retina. The function of the lens is to have the focal point on the plane of the retina, so that the image will be clear. In the case of nearsightedness, the eyeball is longer than the focal length. The image is focused in front of the

plane of the retina. In farsightedness, the eyeball is shorter than the focal length. The image is focused behind the plane of the retina. These defects can be corrected by inserting a corrective lens between the image and the lens.

9. *On the bell-shaped rim of the jellyfish shown in figure 36.27 are tiny saclike structures that hold calcium particles next to a sensory cilium. When the bell tilts, the particles slide over to the cilium. Would you assume that these structures contribute to the sense of taste, smell, balance, hearing, or vision.* 36.4 These calcium particles function as statoliths, used in the sense of balance. When the animal tilts to one side, the particles will move to a different part of the sensory cilium, which will provide information about direction and orientation of movements.

CHAPTER 37

ENDOCRINE CONTROL

1. *Name the endocrine glands that occur in most vertebrates and state where each is located in the human body.* 37.1 Pituitary gland: at the base of the brain. Pineal gland: at the base of the brain, but posterior to the pituitary. Thyroid: in the neck on each side of the trachea. Parathyroid: four glands embedded in the neck posterior to the thyroid. Adrenals: two on top of each kidney (cortex on the outside, medulla on the inside). Gonads: testes are in the scrotal sac, ovaries are in the lower peritoneal cavity. Thymus: behind the breastbone and between the lungs. Pancreas: adjacent to small intestine. Various cells of the gut, liver, kidneys, and placenta also function as endocrine glands.

2. *Distinguish among hormones, neurotransmitters, local signaling molecules, and pheromones.* 37.1 Hormones are chemicals secreted directly into the bloodstream, where they are carried to specific target organs or cells that possess appropriate receptor molecules on their surfaces. The hormones trigger specific responses by the

target cells. A neurotransmitter is a chemical secreted by a neuron and is responsible for the transmission of a nerve impulse across a synapse. The transmitter changes the permeability of the postsynaptic cells. Local signaling molecules are secreted by many cell types. They usually do not enter the bloodstream, and they produce their effects in cells close to their origin. An example is prostaglandin. Pheromones are secreted by exocrine glands and diffuse through the water or air to targets outside the body. These chemicals are used to communicate between members of the same species. The origin, transport, and function of each of these substances are different.

3. *A hormone molecule binds to a receptor on a cell membrane. It does not enter the cell, however. Rather, binding activates a second messenger inside the cell that triggers an amplified response to the hormonal signal. State whether the signaling molecule is a steroid or a peptide hormone.* 37.2 The signaling molecule is a protein that is unable to cross the plasma membrane. A steroid hormone would easily cross the membrane and bind to internal receptors.

4. *Which secretions of the posterior lobe of the pituitary gland have the targets indicated? (Fill in the blanks.)* 37.3 The blanks from left to right are: Antidiuretic hormone and oxytocin. Posterior lobe: Left to right, answers are ADH to kidneys, oxytocin to mammary glands and uterus.

5. *Which secretions of the anterior lobes of the pituitary gland have the targets indicated? (Fill in the blanks .)* 37.3 The blanks from left to right are: Corticotropin (ACTH), Thyrotropin (TSH), Follicle Stimulating Hormone (FSH), Leuteinizing Hormone (LH), and Prolactin (PRL).

CHAPTER 38

PROTECTION, SUPPORT AND MOVEMENT

1. *List the functions of skin, then distinguish between its regions. Is the hypodermis part of skin?* 38.1 The skin protects the body from abrasion, ultraviolet radiation, bacterial attack, and other environmental insults. It is a waterproof barrier. It helps to control internal temperature. It serves as a reservoir for blood. It receives and transmits information about the environment to the brain. It synthesizes vitamin D, which is required for calcium metabolism. Structures derived from the skin include such diverse things as scales, feathers, hair, beaks, hooves, horns, nails, claws, and quills, all of which provide a number of useful functions for those animals that develop them. The word hypodermis means below the skin. It is not part of the skin, but it anchors the skin to underlying structures and allows it to move.

2. *Name four cells types in skin and their functions.* 38.2, 38.3 The four cell types found in the skin are kerotinocytes, melanocytes, Langerhans cells and Granstein cells. The kerotinocytes produce the protein keratin that is incorporated into hair cells and fingernails. The kerotinocytes produce a strong, flexible and cohesive integument because they have adhesions junctions anchored to keratin fibers within them. The melanocytes produce the brownish-black pigment, melanin that protects the body from harmful ultraviolet radiation. Langerhans cells are produced in the bone marrow, migrate to the skin and function as phagocytes that ingest or mark invading pathogens for destruction by the immune system. Granstein cells somehow suppress the action of leukocytes and the immune response.

3. *Distinguish between:*
   a. *Sweat gland and oil gland* 38.2
   b. *Hydrostatic skeleton, exoskeleton, and endoskeleton* 38.4
   c. *Red marrow and yellow marrow* 38.5
   d. *Ligament and tendon* 38.6, 38.7

(a) The sweat glands secrete sweat to assist in cooling the body surface. Oil glands are also called sebaceous glands. They secrete oil that lubricates and softens hair and keeps the skin soft and moist. (b) A hydrostatic skeleton is a closed system containing fluid that can be placed under pressure to produce movement. The tube feet of starfish are excellent examples of a functional hydroskeleton. The exoskeleton is an external skeleton as the prefix exo- denotes. One of the important adaptations of the arthropods is the presence of a hard chitinous exoskeleton that is similar to a suit of armor. The arthropods are vulnerable when they shed their skin during molting. The endoskeleton is an internal (endo- inside) skeletal system such as found in the vertebrates. Muscles are attached to bones to allow movement, (c) The red marrow is found in the lumen of some bones such as the sternum. It is the major site for blood cell formation. Yellow marrow is found in the cavities of most mature bones. It is primarily composed of fat tissue, but it can be converted to functioning red bone marrow if there is a need for large amounts of new red blood cells, (d) Ligaments connect bones to bones at joints. The ligaments of the knee are often subjected to damage in athletic activities. Tendons are sheaths of connective tissues that are found at the origin and insertion of muscles to bones.

4. *What are the functions of bones?* 38.5 Bone tissue functions in movement, protection, support, mineral storage, and blood-cell formation.

5. *Name the three types of muscle, then state the function of each and where they are located in the body.* 38.7 The three types of muscles are skeletal, cardiac, and smooth. The cells of all three muscle types have the following three functions in common: (1) excitability, (2) contractibility, and (3) elasticity. The skeletal muscles are voluntary and responsible for movement of the body. Cardiac muscles are found in the heart and contract to force the blood through the heart and out to the lungs or the body. Both skeletal and cardiac muscles are called striated because of the appearance of myofibrils composed of alternating thick myosin filaments and thin actin filaments. The smooth muscles are found in the body walls of the organs of digestion and circular muscles (sphincters) that control the entrance of material into capillary beds or the stomach and small intestine. In peristalsis the smooth muscles contract rhythmically to mix food with enzymes, and in segmentation the food is moved slowly down the digestive tract.

6. *Look at Figure 38.19 and 38.20. Then, on your own, sketch and label the fine structure of a muscle, down to one of its individual myofibrils. Identify the basic unit of contraction in the myofibrils?* 38.8 Proceed as indicated. The basic unit of contraction of the myofibril are the actin and myosin filaments within one sarcomere.

7. *What role does calcium play in the control of contraction? What role does ATP play, and by what routes does it form?* 38.9 According to the sliding filament theory, the component sarcomeres of the myofibril shorten when the actin filaments physically slide over the myosin filament. For the sliding movement to occur, the heads on the myosin filaments must attach to binding sites on the actin molecules. When attached, the myosin heads are cross-bridged between the two filaments; in a series of power strokes, the myosin heads move from one binding site on the actin to the next in a rachetlike action. The energy needed to detach the myosin heads to allow the sliding of the actin filament comes from ATP. Without ATP, the cross-bridges cannot detach, and the muscle is locked in place, as occurs in rigor mortis. Calcium is stored in the sarcoplasmic reticulum of muscle cells. It is released when a stimulus from a motor neuron triggers an action potential in a muscle cell. The released calcium ions diffuse through the myofibrils reaching the binding sites on the actin. The calcium unblocks the actin binding sites and allows cross-bridging by the myosin filaments to the unblocked binding sites. This allows contraction to occur as the myosin slides by the actin after a new cross-bridge is formed.

CHAPTER 39

CIRCULATION

1. *Define the functions of the circulatory system and the lymphatic system.. Distinguish between blood and interstitial fluid.* 39.1 Humans have a closed circulatory system consisting of a heart (muscular pump), blood vessels (arteries, arterioles, capillaries, venules, and veins), and blood. The circulatory system is designed to transport materials to and from every cell in the

body. The materials that are transported include oxygen, carbon dioxide, glucose, wastes, minerals, vitamins, and hormones. The lymphatic system consists of lymph, capillaries, vessels, and lymph nodes. It has three functions: (1) collects water and plasma proteins and returns them to general circulation, (2) transports fats absorbed in the small intestine, and (3) transports foreign particles and cellular debris to the disposal centers, the lymph nodes.

Blood is a fluid connective tissue composed of water, solutes, and formed elements (red and white blood cells and platelets). It is found in the various blood vessels throughout the body. Interstitial fluid is the portion of the extracellular fluid occupying spaces between cells and tissues.

2. *Describe the cellular components of blood. Describe the plasma portion of blood.* 39.1 The cellular components of the blood consist of three types of cells—erythrocytes (red blood cells), leukocytes (white blood cells), and platelets. The erythrocytes function in oxygen and carbon dioxide transport. There are five types of leukocytes: neutrophils, eosinophils, basophils, lymphocytes, and monocytes. These cells provide defense against invaders. The platelets function in blood clotting. Plasma proteins represents fifty to sixty percent of the total volume of the blood. Most of the plasma consists of water and incorporates the plasma proteins and dissolved substances. The plasma proteins include albumin; fibrinogen; and alpha, beta, and gamma globulin. In addition, the plasma includes various ions, sugars, amino acids, lipids, hormones, vitamins, and dissolved gases.

3. Distinguish between *systemic and pulmonary circuits.* 39.1, 39.2 The systemic circuit includes the left half of the heart (atrium and ventricle), which pumps oxygenated blood to all parts of the body, where it supplies all body tissue with nutrients and oxygen and then returns to the right side of the heart. The pulmonary circuit includes the right half of the heart, which pumps deoxygenated blood to the lungs, where gas exchange takes place, and oxygenated blood flows back to the left atrium of the heart.

4. *State the functions of arteries, arterioles, capillaries, veins, and lymph vessels.* 39.2, 39.5, 39.7, 39.8 Arteries (large blood vessels) carry blood away from the heart and toward the tissues of the body. Arterioles are smaller branches of arteries that continue

to transport blood away from the heart and toward the tissues of the body. Arterioles serve as sites that regulate the direction and amount of blood flow to various parts of the body. Capillaries serve as connectors between arterioles and venules and bring the blood in close contact with individual cells throughout the body. Venules drain the capillaries and bring blood back to the veins leading to the heart. Capillaries serve as diffusion zones to allow for exchange of material between blood and tissue. Veins collect blood from smaller venules and deliver blood to the atria of the heart. The lymph vessels collect the lymph from tissues and conduct it through the lymphatic system to collecting ducts (right lymphatic and thoracic ducts) that empty into veins in the neck region.

5. *Distinguish between the functions of the human heart's atria and ventricles. Then label the heart's components:* 39.6 The function of the two atria is to collect blood from the superior and inferior vena cava (right atrium) and the pulmonary vein(left atrium. When the heart contracts the collected blood will flow from the atria through the atrioventricular valves into the ventricles. The ventricles have larger muscular walls and place the blood under higher pressure to force it to flow to the lungs (pulmonary circuit) and the rest of the body (systemic circuit). The parts of the human heart are labeled in a clockwise direction starting at the top right: (one o'clock position) arch of aorta, trunk of pulmonary artery, left semilunar valve, branches of the left pulmonary vein, left atrium, left A-V valve, left ventricle, endocardium (inner membrane), myocardium (thick cardiac muscle layer), pericardium (outer membrane), septum (partition between the ventricles), inferior vena cava (from trunk, legs), cone-shaped cardiac muscles, right ventricles, right A-V valve, right atrium, right semilunar valve, superior vena cava (from head and arms).

CHAPTER 40

IMMUNITY

1. *While jogging barefoot along a seashore, some of your toes accidentally land on a jellyfish. Soon the toes are swollen, red, and warm to the touch. Describe the events that result in these signs of inflammation.* 40.3 The introduction of a foreign substance (jellyfish material) into the body results in overcoming the body's first line of defense, an intact skin. The introduction of the foreign substance initiates the nonspecific defense reaction known as inflammation, which involves the following five events: (1) blood vessels in damaged or invaded tissues dilate and leak; (2) seepage produces heat, redness, and swelling; (3) an influx of infection-fight proteins occurs; (4) phagocytes migrate to affected tissue; and (5) tissue is repaired. The complement system consisting of about twenty plasma proteins enhances the defense response. Additionally, histamine is secreted into the affected area and enhances the inflammatory response. Chemical gradients are established to attract phagocytotic cells to engulf the invading cells. Interleukins, secreted by active macrophages, induce fever and drowsiness.

2. *Distinguish between:*
   a. *neutrophil and macrophage* 40.3
   b. *cytotoxic T cell and natural killer cell* 40.4
   c. *effector cell and memory cell* 40.4
   d. *antigen and antibody* 40.4

(a) Neutrophils are the most abundant phagocytic white blood cells. Macrophages (meaning "large eaters") are derived from circulating monocytes that act slower than neutrophils and will phagocize a greater variety of substances. (b) Cytotoxic (cell/poison) T cells, as the name implies, are T lymphocytes that kill certain body cells. These infected body cells are identified by the presence of surface antigens and include such cells as those infected with intracellular pathogens, tumor cells, and cells donated in organ transplants. Natural killer (NK) cells arise from stem cells in the bone marrow just as the cytotoxic killer cells do. Their arousal is not dependent on encounters with antigen-MHC complexes, but these cells are triggered to kill tumor cells based on other surface clues. (c) B and T lymphocytes are triggered to divide to form many clones after they come in contact with nonself (foreign) markers. Some subpopulations are designed to act immediately against the marked cells. These are fully differentiated cells known as effector cells. Memory cells are other subpopulations of B and T cells that are inactive. These cells are the basis for the rapid immune response to a second invasion. They "remember" the invader by recognizing external markers and are immediately marshaled into action when a second invasion occurs. (d) Antigens are large molecules, such as foreign proteins, that elicit an immune response by causing the production of specific antibodies. Antibodies are Y-shaped receptor molecules with binding sites for specific antigens. Antibodies are produced by B cells and are the basis of the specific defense response—the immune system.

3. *Describe how a macrophage becomes an antigen-presenting cell.* 40.4 Recall that each pathogenic agent or cell bears specific unique molecular markers that serve as identity markers. In addition, the host organism also bears markers known as self-markers (also called MHC markers) to distinguish them from all other nonself (foreign) markers. Phagocytes begin by engulfing foreign cells, destroying them, and taking individual antigen molecules or fragments of antigens and placing them on the external cellular surfaces. Any cell that processes and displays antigens along with MHC molecules is an antigen-presenting cell. The exposure of virgin cytotoxic cells and virgin helper T cells produces a cascade of new cells that form subpopulations of effector cytotoxic T cells and effector helper T cells to antigen-presenting cells and memory cells. When virgin B cells contact the cells with processed antigen-presenting cells, they will also undergo division to produce clones of memory cells and effector B cells. These cells secrete specific antibodies based upon the antigens on the surface of the antigen-presenting cells. The memory B and T cells are available to initiate the very rapid secondary immune response as soon as a second exposure to the antigen occurs.

4. *Why is a vaccine to control AIDS so elusive?* 40.11 One of the reasons it is difficult to cure AIDS is that the human immunodeficiency virus actually attacks our immune system that has the responsibility for controlling infectious diseases. Secondly,

the virus becomes sequestered inside host cells and inaccessible to antibodies or agents that might be administered to treat the disease. Another feature is the extremely rapid replication of virus particles during certain phases of the infection exceeds the capability of the body to destroy them until the number of helper T cells remaining are insufficient to mount effective immune responses. The virus has an exceptionally high mutation rate that mitigates against developing an effective vaccine or that develops resistance against the drugs that have proven effective.

CHAPTER 41

RESPIRATION

1. *Define respiratory surface. Why must oxygen and carbon dioxide partial pressure gradients across it be steep?* 41.1 A respiratory surface must expose a large surface area to atmospheric gases. The surface must be moist so that oxygen can dissolve and be carried by diffusion to the membranes of the blood cells that will transport the oxygen to the tissues. In the tissues, carbon dioxide is released and carried back to the respiratory surface for simple exchange. Many organisms are equipped with muscles that contract to increase gas flow over the exchange surface. In some others, a countercurrent system is found, in which the blood flow is in an opposite direction to the gas flow to increase the efficiency of gas exchange over the respiratory surface. According to Fick's Law the two controls over the movement of molecules across a respiratory surface are the amount of surface area available and the size of the partial pressure gradient. Thus the greater the partial pressure gradient the fast will be the diffusion (gas exchange).

2. *What is the name of the respiratory pigment in red blood cells? What is the name of a major respiratory pigment in skeletal muscle cells?* 41.1, 41.8 The respiratory pigment in red blood cells is hemoglobin. The respiratory pigment in skeletal muscle cells is myoglobin.

3. *A few of your friends who have not taken a biology course ask you what insect lungs look like. What is your answer (assuming*

your instructor is listening)? 41.2 Aquatic insects breathe through gills. Terrestrial insects have a tracheal system, a system of branching tubes that extend from their outer integuments throughout their bodies. Instead of having a specific organ, such as a lung, where gas exchange takes place, the insects use tracheal tubes that continuously divide, spread in all directions, and dead-end at fluid-filled tips, where gas exchange between the tissue and tracheal system takes place. Since gas with oxygen permeates the insect's internal tissues via the tracheal system, the terrestrial insects do not have a well-developed circulatory system, which otherwise would have to distribute gases throughout the insect's body.

4. *Briefly describe how countercurrent flow through a fish gill is so efficient at taking up dissolved oxygen from water.* 41.3 Countercurrent flow refers to the fact that the fluids in two adjacent areas are flowing in opposite directions. Water moves from the mouth to the pharynx and then over the branched filaments of the gills. Blood vessels adjoin the gill filaments. The inflowing water, rich in oxygen, flows over a blood vessel coming from the body tissue that is low in oxygen and high in carbon dioxide. The first area of contact between the inflowing water and blood vessels is an area of sharp concentration gradient for both oxygen and carbon dioxide. Carbon dioxide rapidly leaves the blood vessel and oxygen leaves the water. As the water continues to move through the gills the exchange rate diminishes but exchange of the gases continue as long as there is a gradient between the water and the blood vessels.

5. *Does the respiratory system of fishes, amphibians, reptiles, birds, or mammals have air sacs that ventilate the lungs?* 41.3 typically five air sacs are attached to the lungs of birds. When the birds inhale air passes over the vascularized respiratory surface in the lungs on its way to the air sacs. When the birds exhale the air again passes over the respiratory surface (lungs) on its way out of the body. Thus, gas exchange takes place during both inhalation and exhalation.

6. *Distinguish between:*
   a. aerobic respiration and respiration C1, 41.1
   b. pharynx and larynx 41.4
   c. bronchiole and bronchus 41.4
   d. pleural sac and alveolar sac 41.4, 41.5

(a) Aerobic respiration is a metabolic pathway that produces enough energy to support an active life-style. It refers to chemical reactions that take place in individual cells. Respiration, on the other hand, is the physical processes involved in gas exchange and refers to muscular contractions associated with inspiration and expiration. (b) Pharynx is the scientific term for the throat. Larynx is the scientific name for the voice box. (c) Bronchioles are the smaller divisions of the bronchus in the respiratory tree. The bronchus is one of the two main divisions of the trachea. Each bronchus is a tube that supplies each lung with its gas supply. (d) The pleural sac is the membrane that surrounds and encases the lungs. The alveolar sacs are the small pouches that surround the grapelike clusters of alveoli that surround the ends of the finest bronchioles.

7. *Explain why humans,(Figure 41.22) cannot survive on their own, for very long, under water.* 41.8 (1) When a person drowns, insufficient oxygen reaches the respiratory surface because the lungs fill with water. (2) The human respiratory system is not designed to allow sufficient water to move by the respiratory exchange surface to supply the body with enough oxygen to maintain life. (3) Water does not have as high a concentration of oxygen as is found in the atmosphere. (4) Many fish exhibit a countercurrent system to increase the efficiency of delivering oxygen from water to the circulatory system. Humans lack this adaptation. For these reasons, a human cannot survive underwater without a supply of gaseous oxygen.

8. *Define the functions of the human respiratory system. In the diagram below, label its components and the major bones and muscles with which it interacts during breathing.* 41.4 The two primary functions of the human respiratory system are obtaining oxygen from the air and removing carbon dioxide. The lungs may also remove other gaseous waste such as alcohol (breathalyzer tests). Starting at the top right (one o'clock position) and proceeding in a clockwise direction. The labels for the diagram are

as follows: pharynx, larynx, epiglottis, trachea, bronchial tree, alveoli, diaphragm, intercostal muscles, lung (one of two), pleural membranes, epiglottis and nasal cavity.

CHAPTER 42

DIGESTION AND HUMAN NUTRITION

1. *Define the five key tasks carried out by a complete digestive system. Then correlate some organs of such a system with the feeding behavior of a particular kind of animal.* 42.1 The five functions of the complete digestive system include: (1) mechanical processing and mobility to break up, mix, and propel food, (2) secretion of enzymes and other substances into the tube, (3) digestion-the chemical breakdown of complex food substances into simple soluble nutrient molecules small enough to be absorbed, (4) absorption-movement of digested substances from the tube through the tube wall to the body, and (5) elimination-the expulsion of undigested and unabsorbed residues at the end of the tube.

2. *Using the diagram on the next page, list the organs and the accessory organs of the human digestive system. On a separate sheet of paper, list the main functions of each.* 42.2; Table 42.4 The organs are described starting at the top right (one o'clock) and proceeding in a clockwise direction: (1) Mouth, or oral cavity—food is chewed to produce smaller particles with increased surface area available to digestive enzymes. Polysaccharide digestion begins here. The food is shaped into a bolus that is easily swallowed. (2) Pharynx—the throat is a passageway from the mouth to the esophagus. It functions by swallowing the food. (3) Esophagus—a tube connecting the mouth with the stomach. Smooth muscles lining the esophagus begin the muscular contractions known as peristalsis, which forces the food down the esophagus and through the cardiac sphincter into the stomach. (4) Stomach—a large muscular sac that mixes food with proteolytic enzymes and other secretions and turns the food into a fluid called chyme. (5) The pancreas functions as an exocrine gland to produce pancreatic juice that is collected and

carried to the small intestine. Pancreatic juice contains sodium bicarbonate, which acts as a buffer to neutralize the acid released by the stomach. It also contains pancreatic amylase, trypsin, chymotrypsin, carboxypeptidase, lipase, and pancreatic nucleases to break down most organic foods in the small intestines. Pancreatic juice is carried from the pancreas to the small intestine by the pancreatic duct. (6) The large intestine stores and concentrates undigested material by absorbing water and salts. (7) The small intestine digests and absorbs most nutrients. (8) The rectum is the last portion of the large intestine. It maintains controls over the elimination of wastes. (9) The opening of the rectum is the anus, where defecation occurs. (10) The gall bladder stores bile produced by the liver. The bile contains no enzymes, but it does emulsify fat to make it easier to digest. The liver secretes bile and bicarbonates to aid in the digestive process. (11) Salivary glands produce salivary amylase that breaks down starch into disaccharide maltose, mucin that moistens the food to allow for ease in swallowing, and bicarbonate to buffer the environment of the mouth.

3. *Name the breakdown products small enough to be absorbed across the intestinal lining , into the internal environment.* 42.5
The four breakdown products are monosaccharides, amino acids, fatty acids, and monoglycerides.

4. *Define segmentation. Does it proceed in the stomach? Does it proceed in the small intestine, colon, or both?* 42.5, 42.7
Segmentation is an oscillating movement induced by the repeated contraction and relaxation of rings of smooth muscles in the walls of the small intestine. This action forces the chyme against the walls of the intestine where digested food is absorbed. Segmentation forces the intestinal contents down the digestive tract. Segmentation continues in the colon but at a slower pace.

CHAPTER 43

THE INTERNAL ENVIRONMENT

1. *State the function of the urinary system in terms of gains and losses for the internal environment. Name the components of the mammalian urinary system and state their functions.* 43.1 The function of the urinary system is to provide a relatively constant internal environment. It is the chief homeostatic organ of the body. It functions to maintain a relatively constant pH, water volume, and dissolved substances and to eliminate metabolic wastes and toxic substances. The intake of water is controlled by thirst receptors in the hypothalamus. When water needs to be conserved, the hypothalamus triggers the secretion of ADH that affects the permeability of the loop of Henle so that more water is reabsorbed and the urine becomes more concentrated. The hormone aldosterone promotes sodium reabsorption by stimulating cells of the distal tubules and collecting ducts to reabsorb sodium faster so less sodium is excreted. When the body has excess sodium, aldosterone secretion is inhibited so that less sodium is reabsorbed and more is excreted. Filtration that occurs in the glomerulus removes water and small solutes. Many of these substances (such as glucose) can be used by the body so that they are reabsorbed in the proximal tubule and returned to the blood. Although 44% of the urea is reabsorbed the remainder is excreted.

The components of the urinary system consists of two kidneys, two ureters, a urinary bladder and a urethra. The kidneys filter water, mineral ions, organic wastes, and other substances out of the blood. They adjust the filtrate's composition and return all but about one percent to the blood. The small remaining fluid and solutes is urine that is transferred via the ureter to a muscular sac, the urinary bladder where the urine is stored. As the bladder fills, stretch receptors in the walls cause smooth muscles in the bladder walls to force the urine through a urethral sphincter to flow out of the body in a process call urination.

2. *Label the component parts of this kidney and nephron.* 43.1
The structures of the kidney starting at the top should be labeled as follows: renal capsule, kidney cortex, kidney medulla, renal pelvis, ureter. The labels for the nephron from the one o'clock position and preceding clockwise are as follows: start of distal tubule, loop of Henle, proximal tubule, and glomerulus.

3. *Define filtration, tubular reabsorption, and secretion. How does urine formation help maintain the internal environment?* 43.2
Filtration: the process by which blood pressure forces water and solutes out of the capillaries of the glomerulus into Bowman's capsule at the beginning of a nephron. Tubular reabsorption: the diffusion or active transport of water and usable solutes out of the tubular parts of the nephron into interstitial fluid and then back to the peritubular capillaries and into general circulation. Secretion: a regulated stage in urine formation in which ions and other substances move from capillaries into nephrons. Urine formation helps maintain the internal environment by removing the chief metabolic waste, urea in the urine. The urine may also contain excess water and solutes. The removal of these substances enable the content of the interstitial fluid and blood to remain relatively constant.

4. *Which hormone promotes (a)water conservation, (b)sodium conservation, and (c)thirst behavior?* 43.2 Water reabsorption is controlled by vasopressin, the antidiuretic hormone (ADH) from the posterior pituitary. ADH increases the permeability of the distal tubule and collecting duct of a nephron, causing more water to move into the interstitial fluid and eventually into the bloodstream to increase blood volume. Aldosterone, one of the mineralocorticoids is secreted by the adrenal cortex to increase the permeability of the distal tubule and collecting ducts to sodium so that sodium enters the interstitial fluid and is reclaimed by the body. The trigger for the release of aldosterone is a decrease in the volume of the extracellular fluid. Stretch or pressure receptors signal the juxtaglomerular apparatus to secrete an enzyme called renin. Renin acts upon a protein to produce hormone angiotensin II, which triggers the adrenal cortex to release aldosterone. Thirst receptors are located in the hypothalamus. Signals from the hypothalamus trigger the secretion of ADH.

5. *Name and define the physical processes by which animals gain and lose heat. What are the main physiological responses to cold stress and to heat stress in mammals?* 43.6, 43.7 The three physical processes associated with the transfer of heat energy are: radiation, conduction and convection. Animals exposed to infra red radiation will gain heat. Any animal that has any contact with a solid object warmer that itself will gain heat by conduction or if it comes in contact with a colder object it will loose heat. Convection is not as important source of heat exchange but occasionally it is important as in wind chill temperatures in severe cold environments. The heat budget of ectothermic animals is controlled chiefly by behavior. Basking in the infra red radiation from the sun or conduction from warm surfaces will provide heat while burrowing under the earth will avoid excessive heat gain or loss. In the endothermic animals behavioral responses to heat and cold are augmented by metabolic processes. In addition, there are morphological adaptations that reduce heat loss such as layers of fat insulation plus fur or feathers. The primary responses of mammals to cold temperatures include pilomotor response (hair becomes erect), peripheral vasoconstriction (to retain the warm blood in the body core) and shivering (heat produced by muscular contraction). Some animals are adapted to cold environments by nonshivering heat production by brown adipose tissue. Animals respond to heat by peripheral vasodilation and evaporative heat loss (sweating or panting).

CHAPTER 44

PRINCIPLES OF REPRODUCTION AND DEVELOPMENT

1. *What is the main benefit of sexual reproduction? What are some of its biological costs?* 44.1 The primary advantage of sexual reproduction is the variation produced by genetic recombination. In asexual reproduction the parents and offspring are alike. If two favorable mutations occurred in two different asexually reproducing forms there would be no way for these two favorable traits to

appear together without both the mutations occurring in the same lineage. On the other hand, if two favorable mutations occur in different lineages they can be combined as a result of a single mating.

Asexual reproduction is more efficient than sexual reproduction. In sexual reproduction specialize tissues to produce gametes must be provided. Specialized reproductive structures have to be developed and maintained. Hormone systems must be developed to synchronize gamete formation, sexual readiness, appropriate courtship behaviors and pheromones and suitable parental behavior in two individuals to protect the young during the critical early stages of life. For those species that lay eggs considerable nutriments must be provided by the female parent. In those that bear their young alive there are prolonged risks for the female during the gestation period. After birth one or both parents may invest both time and energy in nurturing and protecting the young.

2. *Define the key events during gamete formation, fertilization, cleavage, gastrulation, and organ formation. At which stages is the frog embryo in the photograph at right?* 44.2 Sperm penetration into the egg triggers fertilization, resulting in chemical and physical changes of polarity in the egg . Once the egg has been fertilized, no other sperm will be able to enter the egg. The most striking change is the development of the gray crescent opposite the point of penetration by the sperm. The gray crescent establishes the developmental axis and symmetry of the developing egg. Cleavage is a period of cell division without accompanying cell growth. The shape, position, and communication between cells are established during cleavage. Gastrulation is a period of active cell movement where new, different gene activities are brought about by the formation of three germ layers: endoderm, ectoderm, and mesoderm. Organ formation involves the separation of certain cells and tissues that together develop and differentiate into specific organs. The photograph shows an embryo at the eight-cell stage.

3. *Does cleavage increase the volume of cytoplasm, the number of cells, or both compared to the zygote?* 44.2 Cleavage is a program of mitotic cell division that increases the number of cells but does not change the volume of the egg cytoplasm. Cell division occurs without growth.

4. *Define blastula. What is a blastocoel?* 44.3 A blastula is a hollow ball of cells that surrounds a fluid-filled cavity called the blastocoel.

5. *Define cell differentiation and morphogenesis.* 44.4 In cell differentiation, populations of genetically equivalent cells give rise to subpopulations of phenotypically different cell types. Different genes are active in different cells, resulting in specialization of different cells. In morphogenesis, different cell types become organized into all the specialized tissues and organs of the body. The two events that serve as the foundation for cell differentiation and morphogenesis are the unequal distribution of cytoplasmic structures to daughter cells at cleavage and cell interactions.

6. *Define cytoplasmic localization. At which stage does it occur? Does it play a larger role in cell determination in insects such as Drosophila than in vertebrates?* 44.3, 44.5, 44.6 According to cytoplasmic localization concept the future of cells that form during cleavage is determined by their location in the embryo because they receive different parts of the zygote's cytoplasm. For example, one region of the cytoplasm may have proteins that will induce specific genes while other parts of the cytoplasm will have different proteins that would induce other genes. Cytoplasmic localization occurs during cleavage but its effects, that is activating genes may occur at any time during embryonic development. The cells of the insect embryo are more motile than in vertebrates. The surface of the insect embryo may be mapped according to the differentiation patterns that will be expressed in the adult. The fate map of the insect embryo indicates the polarity that exists in the egg. Maternal-effect genes specify regulatory proteins that are localized in different parts of the egg cytoplasm. They induce or repress *gap* genes that map out broad regions of the insect body. Different concentrations of *gap* gene products switch on *pair-rule* genes that accumulate in bands (stripes) that correspond to body segments. The *pair-rule* gene products activate *segment polarity* genes that divide the embryo into segment-size units. The interaction between these three types of genes control the expression of another class of genes-homeotic genes. The *homeotic* genes control the developmental fate of each segment.

131

7. *Define pattern formation. Then explain the key points of the theory of pattern formation.* 44.6 Pattern formation refers to the specialization of tissues and their orderly positioning in space and their development within the embryo. The key points of the theory of pattern formation are: (1) Master genes are activated in orderly sequence at prescribed times, (2)regulatory proteins influence interactions among the master genes resulting in the appearance of different gene products that are spatially oriented relative to each other in the embryo. (3) Different genes are induced (turned on) and repressed (turned off) along the embryo's anterior-posterior axis and dorsal-ventral axis. Selective gene activation produces chemical gradients of proteins throughout the embryo that help to define each cell's identity. (4) Some master genes known as homeotic genes specify the development of specific body parts. The result of this differential behavior is to inform embryonic cells where they are and what they are to become, in other words, pattern formation.

CHAPTER 45

HUMAN REPRODUCTION AND DEVELOPMENT

1. *Name the two primary reproductive organs of the human male and where sperm formation starts inside them. Does semen consist of sperm, glandular secretions, or both?* 45.1 The two primary reproductive organs of the male are the pair of testes-the sperm producing organ. The penis is part of the male genitalia and is the copulatory organ, but not the sex organ. The sex organ is that tissue in the gonad that produces the gametes - the testes. Sperm formation begins in the seminiferous tissues, coiled tubes found in the testes. Semen is the material released during male ejaculation. The semen consists of sperm and accessory seminal fluid produced by three glands: a single prostate gland, two bulbourethral glands (Cowper's gland) and two seminal vesicles.

2. *Does sperm formation require mitosis, meiosis, or both.* 45.2
Inside the seminiferous tubule undifferential cells called
spermatogonia are forced away from the wall into the lumen where
they undergo mitosis. The resulting cells are diploid and are called
primary spermatocytes. They undergo meiosis to produce haploid
spermatids that will eventually develop into mature sperm. Sperm
formation requires both mitosis and meiosis.

3. *Study figure 45.5. Then, on your own sketch the feedback loops
to the hypothalamus and anterior pituitary from the testes that
govern sperm formation. Include the names of the releasing
hormone, hormones, and cells in the testes involved in these loops.*
45.2 As the level of testosterone drops the hypothalamus secrets
gonad releasing hormone (GnRH). GnRH stimulates the anterior
pituitary to secrete the luteinizing hormone (LH) and the follicle
stimulating hormone (FSH). LH triggers the Leydig cells in the
testes to produce and release testosterone. Sertoli cells bind FSH
and testosterone (and function in spermatogenesis at puberty).
Testosterone stimulates the production and development of sperm.
High sperm count causes Sertoli cells to secrete inhibin which
inhibits the secretion of GnRH and LH. Elevated levels of
testosterone inhibits the secretion of GnRH.

4. *Name the two primary reproductive organs of the human
female. What is the endometrium?* 45.3 The two primary
reproductive organ of the female are the ovaries. They produce the
female gametes (eggs). The endometrium forms the inner wall of
the uterus. It covers the muscular layer, the myometrium. It
participates in the formation of the placenta.

5. *Distinguish between:*
   a. *oocyte and ovum* 45.3, 45.6
   b. *follicular phase and luteal phase of menstrual cycle* 45.4
   c. *follicle and corpus luteum* 45.4
   d. *ovulation and implantation* 45.3, 45.7

(a) The primary oocyte is found inside the follicle. Shortly before
the follicle ruptures, the primary oocyte undergoes chemical and
cytological changes, including the completion of meiosis I to
produce the secondary oocyte and the first polar body. The ovum is
the oocyte that will complete meiosis II only after fertilization
has occurred. (b) The menstrual cycle is divided into three phases,

the follicular phase, ovulation, and the luteal phase. The follicular phase covers the first half of the menstrual cycle and involves the breakdown and reconstitution of the endometrial lining of the uterus and the development of the follicles leading to ovulation. The luteal phase involves the conversion of the follicle to the corpus luteum and the preparation of the uterine lining for pregnancy. (c) The follicle is the space where the primary/secondary oocyte develops. The corpus luteum (yellow body) is produced inside the follicle after ovulation. It will enlarge and secrete progesterone. If the egg is fertilized and implanted, the corpus luteum will be maintained, and if fertilization does not occur, it will disintegrate.

(d) Ovulation is the rupture of a mature follicle with the release of an ovum. Implantation occurs two to three days after fertilization. The egg is fertilized in the upper reaches of the female reproductive tract. The zygote starts cleavage as it is impelled down the oviducts by currents created by the ciliated epithelial cells lining the lumen of the oviducts. As the zygote travels, cleavage is initiated. The development reaches the blastocyst stage by the time it reaches the uterus. The blastocyst is implanted on the endometrial wall of the uterus.

6. *What is the menstrual cycle? Name four of the hormones that influence this cycle. Which one triggers ovulation?* 45.3, 45.4 The menstrual cycle is the human counterpart of the typical mammalian estrus cycle. The menstruation cycle involves the development of a follicle, ovulation, and the preparation of the endometrial lining of the uterus for implantation. If the egg is fertilized, the zygote starts cleavage as it travels down the oviduct to implant on the uterine wall as the blastocyst. If the egg is fertilized, the follicle develops into the corpus luteum. The corpus luteum produces progesterone, and the developing placenta secretes chorionic gonadotropin and more progesterone. These hormones prevent the secretion of FSH so that no new follicles develop. If the egg is not fertilized, the corpus luteum fails, and the uterine lining is sloughed off in the menstrual flow. The normal cycle takes about twenty-eight days. The four hormones that influence the cycle are FSH (follicle stimulating hormone), LH (luteinizing hormone), estrogen and progesterone.

7. *Study figure 45.8. Then, on your own, sketch the feedback loops to the hypothalamus and anterior pituitary from the ovaries that govern the menstrual cycle. Include the names of the releasing hormone, hormones, and ovarian structures involved in these loops.* 45.4 Gonadotrophic releasing hormone (GrRH) stimulates the anterior lobe of the pituitary to release the follicle stimulating hormone (FSH) and the luteinizing hormone (LH). FSH and LH stimulate the growth of a follicle, maturation of an oocyte, production and secretion of estrogen and preparing the endometrium for implantation. The increase in the level of estrogen in the blood will trigger a surge in LH secretion by the anterior lobe of the pituitary. The surge in LH produces ovulation and initiates the development of the corpus luteum in the abandoned follicle. The corpus luteum secretes progesterone and estrogen that will maintain the endometrial development if pregnancy occurs. The rise in progesterone and estrogen levels in the blood will suppress the production of FSH and LH and prevent the development of a new follicle.

8. *Describe the cyclic changes that occur in the endometrium during the menstrual cycle. What role does the corpus luteum play in the changes?* 45.4 In the first five days of the menstrual cycle, the endometrial lining breaks down and is sloughed off in the menstrual flow. Under the influence of FSH there is a rebuilding of the endometrial lining with increased vascularization. The endometrium is ready for the implantation of the fertilized egg. If fertilization does not occur, the cycle begins again. If fertilization occurs, the corpus luteum develops and secretes progesterone and estrogen further enhancing the development of the endometrium. Progesterone prepares the lining for the arrival of the blastocyst. The increased levels of progesterone and estrogen suppress the secretion of FSH and LH which prevents the development of other follicles. If the blastocyst does not implant the corpus luteum will secrete prostaglandins and self destruct. When the progesterone and estrogen levels in the blood drop after the demise of the corpus luteum the pituitary will begin to secrete FSH and LH again. The endometrial lining, lacking the stimulus of the secretions from the corpus luteum begin to breakdown and slough off. The menstrual flow will last three to six days. The higher levels of FSH and LH in the blood will start a new menstrual cycle.

9. *Distinguish between the embryonic period and fetal period of human development. Then describe the general organization of a human blastocyst.* 45.7 An average pregnancy lasts 38 weeks from the time of fertilization. It takes about two weeks for the blastocyst to form. From the third to the eighth week is called the embryonic period. This is the period of organogenesis. From the ninth week onward is the fetal period during which organs enlarge and become specialized. Growth and development continue throughout the fetal period and after birth.

10. *State the embryonic source of the amnion, yolk sac, chorion, and allantois. Then state the role that each extraembryonic membrane plays in the structure or functioning of the developing individual.* 45.7 The amniotic cavity arises between the embryonic disk and the blastocyst surface. The amnion is a fluid filled cavity that surrounds and protects the developing embryo. It serves as a shock protector and a thermal insulator and a fluid filled environment where the embryo can move freely. The yolk sac is formed by cells that migrate around the inner wall of the blastocyst. In most animals, the yolk sac is the site for the nutritive yolk that serves as a food supply in the land egg. In humans it is a site for blood cell and germ cell formation. During implantation another cavity develops around the amnion and yolk sac called the chorion. The lining of the chorion begins to develop fingerlike projections that intermesh with the highly vascularized endometrium. These villi will become part of the placenta. The third membrane, allantois, forms from an outpouching of the yolk sac. In the land egg of reptiles, birds and other mammals it serves in respiration and waste storage. In humans it serves as a urinary bladder and for blood formation.

11. *Name some of the early organs that are the hallmark of the embryonic period of humans and other vertebrates.* 45.8 Some of the important features of the embryonic period of vertebrates include the primitive streak, neural folds, neural tube, the allantois and yolk sac, somites (paired segments that will give rise to bones and muscles), pharyngeal arches and features of the face, neck, nose, mouth, larynx and pharynx.

12. *Describe the placenta's structure and function. Do maternal and fetal bloodstreams grossly intermingle in this organ?* 45.9 Blood vessels extend from the fetus through the umbilical cord into the

chorionic villi that are dispersed throughout the placenta. The placenta represents an interweaving of vascular tissue of the developing embryo with its mother. The circulatory system of both individuals are intertwined but maintain their separate identities. This is the site where the embryo receives nutrients and oxygen and exchanges carbon dioxide and wastes. There is a placental barrier so that blood supplies do not normally intermix and there is no gross intermingling of the two bloodstreams.

13. *Name the releasing hormone with a key role in labor. Then state the role of estrogen, progesterone, prolactin, and oxytocin in assuring that the newborn will have an ongoing supply of milk.* 45.1-45.2 Oxytocin triggers an increase in the contraction of the myometrium (muscular layer of the uterine wall). During pregnancy estrogen and progesterone promote an increase in the size of the breast and development of the milk production system. Prolactin is a hormone secreted by the anterior pituitary that induces the synthesis of the enzymes needed for milk production. Oxytocin is secreted by the pituitary in response to a baby suckling a breast. Oxytocin causes contractions that force milk into the breast tissue ducts. All four hormones participate in the process of production and release of milk from the breast.

14. *Label the components of the human male reproductive systems and state their functions:* 45.1-45.2 The labels starting from the one o'clock position and proceeding clockwise are as follows. Prostate gland, bulbourethral gland, urethra, penis, epididymis, testis, vas deferens and seminal vesicles. The three glands that produce seminal fluid are the prostate gland, two bulbourethral gland and two seminal vesicles. The urethra is the common tract for the urinary and reproductive system. The penis is the male copulatory organ to insert the sperm into the female reproductive system. The epididymis is the site for storage and maturation of sperm. The testis is the male sex organ that produces sperm and testosterone. The vas deferens is a duct that transfers sperm from the seminiferous tubules to the epididymis. Surgically cutting this tube is called a vasectomy and results in male sterility because the sperm are unable to leave the reproductive tract.

15. *Label the components of the human female reproductive systems and state their functions.* 45.1-45.2, 45.3 The labels starting at the one o'clock position and proceeding in a clockwise

direction are as follows: oviduct, uterus, vagina, ovary. The oviduct, sometimes called the Fallopian tubes, connects the ovary and the uterus. It is the channel for the transport of the egg to the uterus. Tubal ligation involves cutting and tying this tube and results in sterility of the female because the egg is unable to meet the sperm . The uterus is a muscular sac that contains, supports and nurtures the developing embryo. The vagina is the part of the female reproductive system that receives the sperm and forms the birth canal. The ovary is the female gonad that produces the egg and the hormones estrogen and progesterone.

UNIT VII ECOLOGY AND BEHAVIOR

CHAPTER 46

POPULATION ECOLOGY

1. *Define population size, population density, and population distribution. Describe a typical population in terms of several categories for its age structure.* 46.1 Population size simply refers to the number of individuals that share the same gene pool. Population density refers to the number of individuals in a given area (i.e. square meters of land, cubic centimeters or liters of lake water, etc.). Population distribution refers to the general pattern in which members of the population are spread throughout the habitat. Age structure refers to the number of individuals in each of several age categories such as pre-reproductive, reproductive, and post-reproductive ages. The pre-reproductive and reproductive categories form the reproductive base for the population.

2. *Define exponential growth. Be sure to state what goes on in the age category that underlies its occurrence.* 46.2 If there are no limitations imposed on a population, there will be a percentage of the population that will reproduce during a unit of time. The factor is called the biotic or reproductive potential. As more and more organisms reproduce, the number available to reproduce gets larger and larger. As long as more offspring are produced than die, growth of the population will occur. This brings about exponential growth which will continue indefinitely until some limiting factor slows down the growth. The limiting factor may be space, food supplies, or toxic substances produced so that the population will not grow indefinitely. The more individuals in the prereproductive age category the greater the potential exists for population growth when this group reaches reproductive age.

3. *Define carrying capacity, then describe its effect as evidenced by a logistic growth pattern.* 46.3 The carrying capacity of a particular environment is the suitability of that environment to maintain a population of a particular species at equilibrium. As the population grows, there are different environmental factors that will become limiting and dampen the growth of the population producing a gentle s-shaped curve which levels off at the carrying capacity. A low-density population exhibits the typical sigmoid (s-shaped) growth curve with a slow rate of increase at first, which accelerates to a maximum, and then growth slows down and levels off at the carrying capacity of the environment.

4. *Give examples of the limiting factors that come into play when a population of mammals (for example, rabbits or humans) reaches very high density.* 46.3, 46.4 As the density of a population increases, almost any factor needed by the population can become limiting. Common examples include scarcity of food, presence of predators, lack of space and accumulation of wastes or toxins. In some circumstances it could be the lack of pure water or the incidence of disease and parasites. The explosive growth of rabbits in Australia is a classic example of a population explosion. The human population explosion is thought to be the most serious problem in the world. We are currently experiencing many areas of stress that will impose restrictions on population size. These are examples of limiting factors on population growth.

5. *Define doubling time. At present growth rates, how long will it be before the human population reaches 10 billion?* 46.2, 46.6 The doubling time, or the length of time it takes for the size of a population to double, can be determined by dividing the annual rate of growth into 72. For example, if you had $1,000 invested at six percent annual yield it would take 12 years to have the account double to $2,000 (72 divided by 6=12). At present we have a growth rate of 1.55 percent with a world population of 5.8 billion. If the growth rate remains the same, the world population will double to 10 billion in approximately the year 2050.

6. *How did earlier human populations expand steadily into new environments? How did they increase the carrying capacity in. their habitats? Have they avoided some limiting factors on population growth? Or is the avoidance an illusion?* 46.6

Learning, language, and memory enabled man to expand his range much faster than other species. The shift from a hunting-gathering society to an agrarian culture led to many advances in agriculture to allow larger food production (irrigation, fertilizers, machinery, pesticides) and enabled the human population to expand numerically. The germ theory of disease and the development of public-health programs removed many of the causes of death, enabling people to live longer and to expand the size of the human population. People still starve to death. Have we avoided limiting factors? Yes and no. We have eliminated many causes of death, but other limiting factors have taken their place.

CHAPTER 47

COMMUNITY INTERACTIONS

1. *What is the difference between the habitat and the niche of a species? Why is it difficult to define "the human habitat"?* 47.1 A niche is the full range of abiotic and biotic conditions under which a species can live and reproduce. A niche describes how an organism lives and functions rather than a place where an organism lives. The habitat is the place you would go if you were given the assignment to collect a specific plant or animal. The human habitat would be hard to define because man has invaded all environments and extends his presence to all areas. In addition, man modifies his environment to such an extent that it would be impossible to define a specific habitat as being characteristic of man.

2. *Describe competitive exclusion. How might two species that compete for the same resource coexist?* 47.3 The competitive exclusion principle states that no two species can occupy the same niche for any length of time. One of the species will be better adapted and eventually displace the other species. Two organisms that compete for the same resource will have to find some way to partition the niche by feeding at different times or in different places or under different conditions. This is known as resource partitioning. If there is not some way to reduce the direct

competition, one or both species may become extinct whenever they are found in the same place. For example, squirrels and chipmunks compete for many of the same things (such as food, nesting space, and nesting material), and share the same predators. Yet they generally do not directly conflict with one another. Many species of birds have similar niches but are able to specialize or develop certain facets so that they survive. Perhaps one of the best examples is Darwin's finches that developed adaptations (modification of beaks) so that a number of different species evolved on the Galapagos Islands. One important consideration is that the competition on the island was less severe than it would have been on the mainland. The lack of woodpeckers on the island probably explains why the woodpecker finches were able to develop there.

3. *Define the difference between a predator and a parasite.* 47.4 A predator is an organism that feeds on and may or may not kill other living organisms (its prey). Unlike parasites, predators do not live on or in their prey.

4. *Define primary and secondary succession.* 47.7 Primary succession involves invasion of an uninhabited space, such as a lava flow, a new sand dune, or an island forming in the ocean. Secondary succession occurs on abandoned farms, or at sites that have been devastated by a catastrophe such as a fire, hurricane, or explosion.

5. *What is a climax community, and how does the climax-pattern model help explain its structure?* 47.7 A climax community is filled with species that will continue to replace themselves indefinitely. They represent the organisms that are best adapted for that particular habitat.

The organisms in the climax community are in equilibrium with themselves and with the environment. The climax pattern model also provides some flexibility in the composition and characteristics of a climax community. Instead of every climax community having to be identical to others, the climax community is adapted to a pattern of environmental factors, including climate, soil, topography, species interaction, and stresses and recurring disturbances such as flooding or fire. Thus there is a continuum of climax stages to succession. They bear considerable similarity, but there is a possibility for variation in species composition and character without completely changing the climax concept.

CHAPTER 48

ECOSYSTEMS

1. *Define an ecosystem and its trophic levels.* 48.1 "Ecosystem" is short for "ecological system." It is the functional unit of nature. It consists of producers, consumers, decomposers, and raw material. It is a system of living organisms that interacts with the physical and biotic aspects of its environment. Trophic level refers to feeding levels. In the food chain, each organism may represent a different trophic level upon which the next organism in the food chain feeds. The autotrophs or the producers represent the first level. The second trophic level involves the primary consumers known as herbivores, feed upon the producers. The third trophic level includes the secondary consumers, known as carnivores, feed upon herbivores. Still another carnivore may feed on the proceeding one representing another trophic level, in this case, a tertiary consumer. The ultimate source of energy is the sun, but the specific source for the individual is the organism it eats—a member of the trophic level below it. An example of a food chain is presented below:

Sun
Autotroph	(producer)	grass
Primary consumer	(herbivore)	grasshopper
Secondary consumer	(primary carnivore)	bird
Tertiary consumer	(secondary carnivore)	snake
Quarternary consumer	(tertiary carnivore)	hawk

2. *Characterize grazing and detrital food webs.* 48.2 A grazing food web is much more complex than a simple food chain. Not only are there more organisms involved, but also one organism can operate on one trophic level or another depending upon what organism it eats. A grazing food web is characterized by

consumption by herbivores, which are in turn fed upon by carnivores. In a detrital food web, the primary net productivity is used by decomposers and detrivores (not herbivores). The energy captured by the producers flows through both food webs and eventually leaves the system as a result of metabolic activities (aerobic respiration mostly). The energy is lost primarily as low-grade heat.

3. *Describe the reservoirs and organisms involved in one of the biogeochemical cycles* . 48.4 In a biogeochemical cycle, ions or molecules of a nutrient are transferred from an organism to their environment and back to other organisms. There are different reservoirs where these elements are found—inside living organisms, the soil, water and atmosphere (biosphere, lithosphere, hydrosphere and atmosphere). There are three different types of cycles. Hydrological cycle involves the movement between water and organisms in a cycle. Water serves as a reservoir for these elements. In the atmospheric cycle, elements move from organisms to the atmosphere and back again. The atmosphere acts as a reservoir. In the sedimentary cycle, the nutrient travels from the soil to organisms and back again. The soil serves as the reservoir. In these three cycles, the element in the reservoir has a different physical state: liquid, gaseous and solid. The term "biogeochemical cycle" involves the movement of a chemical element through the living organisms (bio), the earth or soil substrate (geo), and any of the chemical compounds it may form (chemical).

4. *Define and describe the connections among nitrogen fixation, nitrification, ammonification, and denitrification.* 48.8 Nitrogen fixation is the reduction of atmospheric nitrogen gas to form ammonia (remember reduction involves the addition of hydrogen, thus $N_2 + H_2 \text{---} > NH_3$). Often this is accomplished by nitrogen-fixing bacteria. Nitrification converts ammonia into various nitrite and nitrate compounds. This process is often accomplished by nitrifying bacteria. In ammonification, nitrogen-containing wastes, various organic compounds, and the bodies of dead plants and animals are decomposed by bacteria and fungi to form ammonia. A bacterial process known as denitrification occurs when nitrates and nitrites are converted to nitrogen gas or nitrous oxide.

CHAPTER 49

THE BIOSPHERE

1. *Define atmosphere, lithosphere, and hydrosphere.* C1 The atmosphere is the gaseous envelope surrounding the earth. The lithosphere includes the outer rocky layer of the earth with the accompanying soils and sediments. The hydrosphere encompasses the waters of the earth in frozen or liquid forms such as the polar ice caps, glaciers, icebergs, oceans, lakes, and running water. These three areas are the areas where life may exist and form the earth's biosphere.

2. *List the major interacting factors that influence climate.* 49.1 The four interacting factors that influence the climate are: (1) the amount of insolation (incoming solar radiation), (2) the daily and annual movement of the earth, (3) distribution of continents and oceans, and (4) elevation. The interaction of the four variables produces prevailing winds and ocean currents. Climate affects the composition and character of soils, which in turn influence the growth and distribution of vegetation, which control the number and distribution of animals. In other words, the climate controls the biological, chemical, and physical characteristics of an ecosystem.

3. *List some of the ways in which air currents, ocean currents, or both may influence the ecosystem in which you live.* 49.1, 49.2 The prevailing air and ocean currents dictate the distribution of ecosystems, primarily through their effects on climate. For example, in the temperate region (north and south latitudes), the prevailing westerlies result in the movement of air masses and their associated weather from west to east. The gyres of the ocean produce characteristic warm and cold currents that affect climatic conditions on land, so that the Gulf Stream makes the British Islands warmer than you would expect. Note also in the appropriate figure in this chapter that the western edge of each continent is characterized by cold currents. These currents promote upwelling, thereby bringing to the surface nutrients that increase the productivity in these areas. The proximity of water tends to

ameliorate weather conditions so that coastal climates are less severe than areas deep in the interior of continents. The distribution of biomes across the world is dependent on the responses of the organisms in the biomes to the prevailing climate. For example, grasslands are dependent upon receiving enough rain. The same may be said for tropical rain forests. Reductions in rainfall could result in destruction of that biome and its transformation to a desert. Make the appropriate observations about the ecosystem in which you live to answer the last part of this question.

4. *Indicate where the following types of biomes tend to be located around the world. Then describe some of their defining features.* 49.3, 49.5-49.9

a	deserts	e. deciduous broadleaf forests
b.	dry shrublands	f. coniferous forests
c.	grasslands	g. alpine tundra
d.	evergreen broadleaf forests	h. arctic tundra

(a) Deserts are located in the rain shadows of tall mountains at thirty degrees north and south latitudes in the northwestern United States, northern Chile, Australia, northern and southern Africa and Arabia, the Gobi of Asia and the Kyzyi-Kum. They occur in land regions where the potential for evaporation greatly exceeds rainfall. Rainfalls usually amount to less then ten inches (25 cm/per year). (b) Dry shrublands and dry woodlands are found in western or southern coastal regions between thirty and forty degrees latitude. They experience less that twenty-five to sixty cm. of rain. The rain occurs during mild winters and the summers are hot and dry. Two shrublands include fynbos near the Mediterranean and the chaparral in southern California. The areas often are set on fire by lighting often followed later by mudslides. (c) Grasslands are found in the interior of continents. In North America, the short grass prairies are found to the west where there is less moisture while the long grass prairie is found to the east where more moisture is available to support the larger grasses. Other major grassland include the African Veldt, the Russian steppes and the pampas of Argentina. The annual rainfall is between twenty-five and one hundred cm which is enough rainfall to prevent the formation of deserts and not enough moisture to support forest. Viparian vegetation including trees will be found along stream banks. In some grasslands, called savannahs scattered trees occur.

146

The rainfall may range from 90 to 150 cm/year with more trees found where there is more precipitation. Often a monsoon wind pattern results in prolong periods of drought followed by rain when the wind shifts so that it comes in from the ocean. (d) Evergreen broadleaf forests cover tropical zones of Africa, East India and Malayan Archipelago, southeastern Asia, South American and Central America. Rainfall amounts to a minimum of 130 cm and may exceed 200 cm/year. These forests dominate the area between twenty degrees north and south latitude. (e) Temperate deciduous forests are found in the temperate zone where winters get cold and leaves are shed in the cold weather. (f) Coniferous forests are found across North America, Asia, and Europe. They are called the taiga, the swamp forests. The rain occurs in the summer. Needle-like leaves of the trees are adaptations for periods of drought. (g) The alpine tundra is found above the timberline in tall mountains throughout the world. Most plants are small and herbaceous although some small dwarf trees are found. (h) The Arctic tundra occurs north of the Arctic circle. The soil is characterized by frozen ice under the ground. It may be over 500 meters thick in some areas and does not melt even during the summer. Since the ice is below ground there is no way for the water to percolate through the ice so that it remains waterlogged. The vegetation is similar to that found in the alpine tundra.

5. *Define soil, then explain how does composition of regional soils affects ecosystem distribution?* 49.4 Five factors affect soil formation. They are: parent material (type of underlying rock), climate, topography, vegetation, and the length of time the factors have been operating. Of course, the climate would vary in different geographic locations. Desert conditions produce entirely different soil types from those soils with high rainfall and leaching or removal of nutrients, as occurs in tropical rain forests. A mild climate supports a temperate deciduous forest, which supports a different type of soil than an evergreen forest in a moist, cool climate that has acidic litter and humus. The color and textures of the soils along with their mineral contents will tend to differ from one climatic region to another. The availability of minerals and amounts of organic material will differ.

The climate, topography, and parent material will influence the vegetation that becomes established and survives in any given area. Grasslands are characterized by rich layers of humus, while in the tropical forests the majority of nutrients are located within the plants, and very little is present in the soil. Many deserts are alkaline flats, while the taiga and tropical rain forests may be acidic.

6. *Describe the littoral, limnetic, and profundal zones of a large temperate lake in terms of seasonal primary productivity.* 49.10 The littoral zone extends from the edge of a lake to the depth at which aquatic plants stop growing. The profundal zone is found below the depth at which light penetrates. The limnetic zone is found between these two areas. Primary productivity is correlated to the temperature. There are two seasonal overturns in the lake brought about through changes in the density of water, the spring and fall overturn. The surface water and deeper water exchange places, bringing nutrients to the surface resulting in a bloom of phytoplankton. There is a dramatic rise in primary productivity as long as the availability of nutrients fuel the increase in productivity. The primary productivity is high in the littoral and limnetic zones while the lack of light precludes the profundal zone from participating.

7. *Define the two major provinces of the world ocean.* 49.11 The ocean provinces are the benthic and pelagic provinces. The benthic province includes all the sediments and rocks of the ocean's bottom. The pelagic province is the entire volume of ocean water.

8. *What points favor the hypothesis that life may not_have originated in the surface waters of the seas?* 49.11 It is logical to think that the conditions at the hydrothermal vents on the ocean floor have been in existence probably much longer than the other environments that we associate with the beginning of life. A constant source of energy and appropriate chemical elements were available for chemosynthetic organisms to evolve and utilize these resources. The sites were certainly protected from lethal ultraviolet radiation. The organisms had to develop features that would enable them to survive the great pressure and the presence of toxic sulfur compounds. Although the habitat is turbulent, it is a place where life could flourish, as witnessed by the success of the vent organisms.

9. *Define and characterize the following ecosystems.* 49.12, 49.13
   a. *estuary*   c.   *sandy shore*
   b. *rocky shore*   e.   *coral reef*

The estuary ecosystems occur along coasts where fresh and saltwater mix. They are some of the most productive and polluted ecosystems in the world. They serve as the reproductive energy base and energy source for much of the life in the ocean. The intertidal zones of rocky and sandy shores are influenced by the character of the substrate. The rocky shore exhibits sharp zonation supporting three zones. (a) The scarcely populated upper littoral zone is often exposed to the air during the day. (b) The midlittoral zone with algae and mobile invertebrates. (c) The most diverse and protected lower intertidal zone. There is rapid erosion that prevents detritus from accumulating so that grazing food webs predominate. The loose sediments of sand and muddy shores is much more unstable to wave and tidal action with few anchored plants. There are few grazing food webs. Detrital food webs supported by organic debris from offshore or washed from the landforms predominate. (d) The coral reef ecosystem is dominated by the wave-resistant masses of coral. They physically protect organisms and provide habitats for extremely high diversity of vertebrate and invertebrate life.

10. *Describe an ENSO event and some of its consequences in both hemispheres.* 49.13, 49.14 An ENSO event refers to an *El Nino* Southern Oscillation. ENSO results from a recurring seesaw in the atmospheric pressure and prevailing currents in the South Pacific Ocean. The massive warm water in the western equatorial Pacific affects the world's air circulation. During an ENSO event the warm air and warm currents move eastward toward the western coast of South America. The cold Peruvian or Humboldt current dies down and the level of the ocean near the coast rises. The changes in water temperature and atmospheric pressure bring about drastic changes in the worlds weather patterns. This periodic phenomenon was only recognized in the early 1970's and total impact on world climates has yet to be fully appreciated or explained. It is thought to be one of the major controlling factors of the weather throughout the world producing droughts and storms.

CHAPTER 50

HUMAN IMPACT ON THE BIOSPHERE

1. *Define pollution and list some specific examples of water pollutants.* 50.1, 50.7 A pollutant is any substance that an ecosystem has no prior evolutionary experience so that adaptive mechanisms are not in place. These substances may be naturally occurring substances such as methane or carbon dioxide that are common greenhouse gases or excessive nitrogen or phosphorus in aquatic ecosystem that leads to eutrophication. They may be chemicals synthesized by man such as chloroflurocarbons that destroy ozone in the upper atmosphere. Ozone may also function as a pollutant in the troposphere. Some examples of water pollutants include the heavy metals that can produce damage through bioaccumulation. Agricultural runoff and sewage from septic tanks or other sources can lead to eutrophication. Thermal pollution near hydroelectric power plants can lead to fish kills. Acid deposition can lead to higher mobility of heavy metals and eventual sterilization of lakes. Many rivers and marine environments in some parts of the world are treated as waste dumps.

2. *Distinguish among the following conditions.* 50.1
   a. industrial smog   *c.   dry acid deposition*
   b. photochemical smog   *d.   wet acid deposition*

(a) Industrial smog produces a gray haze over industrialized sites as a result of burning fossil fuels and the presence of associated pollutants such as smoke, soot, heavy metals and other waste products. (b) Photochemical smoke forms a brown haze over large cities. Transportation wastes, particularly nitric oxide is the primary factor. It combines with oxygen to form nitrous dioxide. When nitrous dioxide is exposed to sunlight it reacts with hydrocarbons to form photochemical smog. Smog is a descriptive term that combines the words smoke and fog. Local physical conditions such as basins surrounded by mountains make these sites more likely areas for smog alerts. Thermal inversions exacerbate these conditions. (c) Dry acid deposition refers to the atmospheric fall out of nitrogen and sulfur compounds under dry conditions.

(d) Wet acid deposition is often called acid rain. It occurs when the nitrogen and sulfur compounds combine with water in precipitation to form nitric and sulfuric acids.

3. *Define CFCs and describe how they apparently contribute to seasonal thinning of the ozone layer in the stratosphere.* 50.2 CFCs are chlorofluorocarbons, compounds that, as their name implies, are composed of chlorine, fluorine, and carbon. They include items such as freon used in air conditioners, foaming agents for packaging materials, industrial solvents, and propellants in aerosol spray cans. When CFCs absorb UV light, they release chlorine, which reacts with triatomic ozone to produce oxygen (two atoms of oxygen) and chlorine monoxide. The chlorine monoxide reacts with free oxygen to release the chlorine, which can then attach to another ozone molecule. One chlorine atom has the potential of destroying 10,000 molecules of ozone to produce oxygen. The depletion of ozone leads to the formation of ozone holes in the atmosphere.

4. *What percent of the Earth's land masses is under cultivation? What percent is available for new cultivation?* 50.3 The human population uses nearly 21 percent of the Earth's land surface for agriculture and grazing. Another 28 percent of the earth could be used for agriculture but its productivity would be extremely limited.

5. *Define and describe possible consequences of deforestation and of desertification.* 50.4, 50.5, 50.6 Deforestation is the removal of trees from large tracts of land to switch to agriculture or grazing. There is a huge demand for wood and forest products. Often overlooked is the world's fuelwood crisis. In the underdeveloped world one important but disappearing resource is wood for fuel (heating and cooking). Demand for hardwood for specialty uses is rapidly exhausting the supply. Pulpwood use in paper production contributes to the problem. Slash and burn have converted parts of the tropical rain forest into land that has marginal and fundamentally limited use for agriculture and grazing. The consequences of deforestation is hard to overestimate. The smoke in the Amazon basin has long been a focal point but the last half of 1997 has also demonstrated a real problem in Southeastern Asia and Malaysia and Indonesia. The burning of forests increases the amount of carbon dioxide in the atmosphere and increases the

threat of global warming. The loss of trees reduces the reservoir of carbon locked in vegetation. Deforestation alters the albedo of the earth's surface so that it reflects more insolation back into space. Deforestation also alters rates of evaporation, transpiration, runoff, erosion and perhaps rainfall patterns. Deforestation has made Madagascar the most eroded country in the world. Removal of trees removes nesting sites and changes environmental conditions to the extent that many species will become extinct and others will become extinct. Deforestation simplifies ecosystems and makes them much more unstable.

Desertification is the conversion of large tracts of grasslands into more desert-like states. Because of population pressures and the desire to produce more animal protein much marginal land has been opened to grazing. The influx of herbivores onto marginal land tends to deteriorate the productivity of the land even further through overgrazing. Changes in weather patterns such as produced by El Nino further promotes new areas where droughts produce dramatic damage to ecosystems. Salination, the build-up of mineral salts after prolonged irrigation, renders some agricultural lands unusable. The spread of deserts is primarily a "patchy" phenomenon. Small areas become dry and devoid of vegetation. In time, the soil becomes less suitable for vegetative growth through erosion and the unproductive land increases and spreads at the margins of the patches.

6. *Which human activity uses the most freshwater?* 50.7 Large-scale agriculture accounts for nearly two-thirds of the human population's use of freshwater.

CHAPTER 51

AN EVOLUTIONARY VIEW OF BEHAVIOR

1. *Explain how genes and their products, including hormones, influence the mechanisms required for forms of behavior.* 51.1 Genes are responsible for the ability of the nervous system to sense,

interpret and respond to the environment. In addition, genes are required to produce hormones that chemically alter an animal's behavior. For example, bird songs are controlled by hormones. Genes produce pheromones which control many aspects of social communication and behavior (mating pheromones, trail pheromones, feeding pheromones).

2. *Define these terms: instinctive behavior, sign stimulus, fixed action pattern, and learned behavior.* 51.1, 51.2 Instinctive behavior differs from learned behavior in that it can be performed without previous experience. Examples include a spider building its web, a dog scratching, the sucking response of a newborn infant. A sign stimulus involves the ability of a newly born or hatched individual to recognize one or two well-defined clues in the environment. An example would be the imprinting of a newly hatched bird to the first large moving object. A fixed action pattern is a simple program of muscular activity that runs to completion after an appropriate stimulus has been provided. A newly hatched cuckoo will force any round object (i.e. an egg) out of the nest. Learned behavior involves processing and integrating information gained from first hand experiences. For example, habituation involves the failure of an animal to respond to a loud noise or the presence of another animal because prior experiences tell it that there is no danger associated with the noise or other animal.

3. *Contrast altruistic behavior with selfish behavior. Why does either form of behavior persist in a population?* 51.3 In altruistic behavior one organism sacrifices itself for its relatives (a member of a grazing herd giving alarm signals when a predator is observed). Selfish behavior provides for individual needs. Both behaviors are advantageous or they would have been eliminated by natural selection. The behavior pattern increases the chances that the individual's genes in selfish behavior, or a relative's genes in altruistic behavior will survive and be passed on to the next generation.

4. *Describe some characteristics of communication signals. Then give an example of a communication display.* 51.4 Communication signals involves cues in the form of specific sounds, odors, colors, patterns, postures, and movements. Female frogs respond to species specific calls of male frogs. Pheromones are responsible for male butterflies finding a receptive female. The pattern of flashes

emitted by a female firefly attracts a male. Threat displays by dogs may lead to conflict or to acquiescent behavior. These patterns are called communication displays. The display of a male peacock's tail feathers a courtship display. Tactile displays are involved in communication in a bee colony to describe the distance and location of a food source.

5. *Describe some feeding behavior or mating behavior in of natural (individual) selection.* 51.1, 51.5 Sexual selection is a form of natural selection operating on individuals. Male birds compete with one another for females. Successful males will pass their genes on to the next generation. The competition is based upon the quality of the song, the appearance of the male and the quality of the territory that he has established in competition with other males. The female's strategy is to select the best of the males thereby insuring that her genes will be coupled with superior male genes and that her contribution to the net generation will be successful. If she makes a poor choice, her chance for contribution to the next generation will be diminished.

6. *List some of the benefits and costs of sociality.* 51.6, 51.7. Some of the benefits of sociality include: cooperative predatory avoidance (by virtue of belonging to a school of fish an individual fish may escape predation), the selfish herd (all members of a herd benefit by giving a mutual alarm signals when a predator is spotted) and dominance hierarchies (provides a system in which members of the group are protected by stronger members of the group). The division of labor in a beehive provides an integrated society in which all members contribute to the good of the hive. The cohabitation of a social group increases the competition between members of the group. Large groups promote the spread of contagious diseases and parasites. Another cost is the risk of being killed or exploited by other members of the group.

7. *Why don't the members of a selfish herd live apart if each one is "trying to take advantage" of the others?* 51.6 The value of large numbers found in a selfish herd is that it provides members of the group protection from predation based upon the concept that there is safety in numbers. If an individual was isolated from a group when a predator was encountered, it would have little hope of escape. The animals in a group try to reach the center, where it is the safest, rather than the unprotected outside.

8.*How do sterile termites propagate their genes?* 51.8, 51.9 Any sterile individual can control the genes that are passed on to the next generation by performing any action that favors its relatives to survive and reproduce. Members at the lower end of the dominance hierarchy (immature males) can help defend the group even though they have not reached the level to reproduce. Worker bees are sterile females, but without their multitude of activities the hive would fail.

In the termite colony, only the queen and a relatively few kings serve as parents for the colony. The remainder including workers and soldiers are sterile. These sterile individuals contribute to the success of the entire colony by their individual behaviors thereby allowing the fertile kings and queen to reproduce other members of the colony. Thus, the genes of the fertile termites are supported and allowed to propagate by the activities of the sterile members of the colony